Celebrating Walter Munk's Centennial

A special edition featuring:

- Greeting by Walter Munk
- Foreword by Carl Wunsch
- Preface by Robert C. Spindel and Peter F. Worcester
- Photographic Update 2010-2017
  by Deborah Day, Mary Coakley Munk, and Brittany Hook
- Tribute by Jim Cairns
- Updated Awards, Honors, and Bibliography

Followed by:

Seventy Years of Exploration in Oceanography

A Prolonged Weekend Discussion with Walter Munk

Hans von Storch • Klaus Hasselmann

Originally published by Springer 2010

MUNK CENTENNIAL

SCRIPPS INSTITUTION OF OCEANOGRAPHY

GLÜCK AUF!

WALTER MUNK'S
100TH

Photo: Eduardo Contreras/San Diego Union-Tribune

As a young boy growing up in Austria, I cherished the time we spent at the Eggelgut, our family home in Altaussee. One of my favorite memories is the traditional greeting, “Glück Auf!” that the salt miners shouted to us, each morning and evening, as they passed by using a short cut through our property, just as they had for hundreds of years. Literally translated, “Glück Auf!” means “Luck Up!”

To my friends and colleagues,
with sincere gratitude and appreciation
for your friendship and support throughout the years.

GLÜCK AUF!

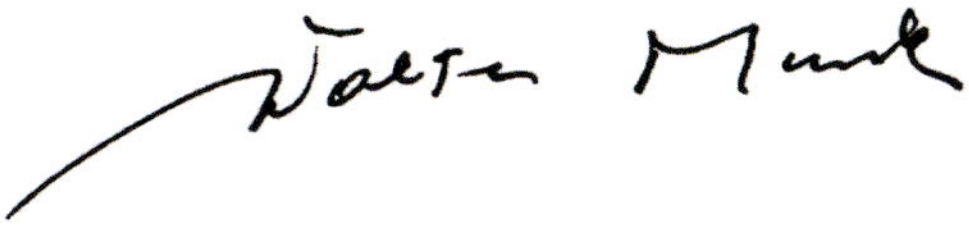

Foreword – Carl Wunsch

Once Again: A Foreword to Walter Munk

What can one say that hasn't already been said? The astonishing Walter Munk is 100 years old, and still carrying on intellectually and socially as though 50 years younger.

Long ago I concluded that Walter was an impractical role model! Physically and mentally. Who could possibly replicate his endless list of scientific contributions? He has the knack of writing a series of notable papers which, with a deceptive simplicity, set entire sub-fields going in new directions, and which decades later are the focus of new ideas. His physical stamina remains astonishing. At the first Symposium in honor of his 100th birthday, Walter sat in the front row for 3 days. He had lunch and celebratory dinners and receptions with the attendees, and as the very last speaker of the Symposium, gave an engaging and interesting talk about his latest work on a major unexplained characteristic of surface gravity waves–his old love from 80 years before. Soon after, he and Mary went to a climate meeting at the U.N. in New York where he spoke. A bit later, they set off on an extended European jaunt.

The two Munk Centennial Symposia have concerned internal waves and mixing, and ocean acoustics. Several more would have been appropriate: Earth rotation; sea-level change; water waves and remote sensing; ocean circulation; the Navy and national security. Most of physical oceanography and much of geophysics have been Walter's domain for the last 80 years–only logistical practicality has limited the breadth of observance.
One of Walter's strongest characteristics is his seemingly endless, unquenchable, optimism. No matter how bleak things are, Walter's attitude has always been that it will all come right, the best people will emerge, and whatever physical problem he is facing will be fixed (four[!] hip replacements; back surgery). The only time I saw Walter professionally discouraged was in the midst of the public hearings about ATOC. People simply wouldn't listen, and Walter's belief that he could persuade anyone of the sense of what he sought was seriously undermined. Obviously, he bounced back and has carried on as before. (I have often thought it was a good thing he did not become a used-car salesman.)

Walter's role as one of the greatest oceanographers and scientists of all time is secure. With the devotion and support of Mary, we all wish him many happy returns!

Note: Some interesting and often amusing essays about Walter can be found in the (mid-career) Festschrift for his 65th birthday (SIO 84-5). An extended oral history with R. E. Doel can be found online at the UC San Diego archive website. Judy Munk used to say, "Walter never let the facts get in the way of a good story!"

Preface on the Occasion of Walter Munk's Centennial – Robert C. Spindel and Peter F. Worcester

Walter H. Munk: Seventy-Five Years of Exploring the Seas†

†This article has been reprinted with permission from *Acoustics Today* (with minor modifications). It was originally published in the Spring 2016 issue.

During a career spanning 75 years, Walter Munk has made seminal contributions to acoustical oceanography, underwater acoustics, physical oceanography, and geophysics.

Some people think there is more than one Walter Munk and with good reason. Acousticians know the Walter Munk who is one of the inventors of ocean acoustic tomography and ocean acoustic thermometry, or the Walter Munk who quantified the effects of ocean internal waves on acoustic propagation, or the Walter Munk of the canonical Munk sound-speed profile. Oceanographers know the Walter Munk who, with Christopher Garrett, devised the Garrett and Munk formulation of the ocean internal wave spectrum (Garrett and Munk, 1972), or the Walter Munk who did pioneering research on ocean tides, tsunamis, and surface waves. Geophysicists know the Walter Munk who studied the Earth's wobble and spin and who, with Gordon McDonald, produced a classic monograph on the subject (Munk and MacDonald, 1960), or they know the Walter Munk who was one of the initiators of the 1962 MOHOLE project to drill into the Earth's mantle.

Walter did not begin to work on underwater sound until the early 1970s after he had made seminal contributions in many other fields. When he was in China in 1978 as the chairman of the first U.S. Oceanographic Delegation, he was asked if it was his father who had worked on ocean waves, or did he have a relative with the same name who worked on tides? When he tried to assure them that both were his early work, they were hard pressed to believe him because, they said, the Walter Munk they knew worked on acoustics! More than one person who has met him has thought that he, Walter Munk, was the son of the Walter Munk. One of us (RS) was a victim of the delusion. RS had used Walter's elegant work (with Chip Cox) on the slope of ocean waves, estimated by observing the sun's glint off the water (Cox and Munk, 1954), in his doctoral research. Like many students reading papers by famous authors, he naturally assumed that the author had long ago passed on. So when RS first met Walter in the mid 1970s he thought he was meeting Walter Munk's son!

Walter H. Munk was born on October 19, 1917, in Vienna, Austria, a country with no coastline, far from the ocean, yet he has spent much of his life in, on, and studying the sea. He was sent to the United States at age 14 to finish high school at an upstate New York preparatory academy. After graduation, he was apprenticed to a financial firm that his grandfather had founded. Apparently, to our good fortune, economics was neither Walter's love nor his forte. He left the firm after three years, and it collapsed a year after that. Then, in his own words: "Somehow I talked my way into Cal Tech and graduated in 1939 in Applied Physics." Because the girl he was seeing at the time vacationed in La Jolla, he took a summer job at the Scripps Institution of Oceanography. Although the romance ended, it was here that his life long passion for the sea was born and ultimately his interest in ocean acoustics. Scripps is where he has spent his entire career.

Figure 1. Scripps Director Harald Sverdrup (left) with Walter on his graduation day, 1946. Photo from Scripps Institution of Oceanography Archives

Scripps was then a small institution with a staff of about 15 led by Harald Sverdrup, the distinguished Norwegian Arctic explorer. Following the summer job, Walter returned to Cal Tech for a Master's Degree, and then the next summer, back at Scripps, he convinced Sverdrup to take him on as a PhD student (Figure 1).

According to Walter, for a time he constituted Scripps' entire student body. He applied for U.S. citizenship when Nazi Germany invaded Austria. Anticipating the outbreak of war, he interrupted his studies, enlisted in the U.S. Army, and spent the next year and a half in the Field Artillery and, because he was an avid and expert skier, in the Ski Troops. One gathers from stories he tells of those eighteen months that neither he nor the Army were unhappy when he was excused from military service in order to pursue defense research at the University of California Division of War Research (UCDWR) with offices at the U.S. Navy Radio and Sound Laboratory in San Diego. Walter says that he was so ineffective as a private that he was promoted to corporal, and he was so good at that that he was promoted back to private.

For the next six years, Walter and colleagues from Scripps at the UCDWR worked primarily on issues related to amphibious landings, starting with developing a forecast system for wave conditions to be expected during the planned Allied landing on the northwest coast of Africa. In winter the coast is subject to a northwesterly swell that produces breakers exceeding six feet on two out of three days. Because the LCVP (Landing Craft, Vehicle, Personnel) tended to broach in seas exceeding six feet, the problem of predicting the right day for the landing was critical to its success. Based almost entirely on empirical data, Sverdrup and Munk developed an analytic representation for a storm-generated, sea-surface spectrum based on wind speed and fetch and then calculated how it would be attenuated beyond the area of the storm and how it would be transformed in shallow areas. The ultimate prediction for the Normandy invasion based on their work was correct—the seas would be rough but would permit landings. This work was published in the now classic H.O. 601 (Sverdrup and Munk, 1947), and the method was used to predict sea conditions for amphibious landings in both the Pacific and Atlantic theaters. A prescient aspect of this work (aside from the fact that it yielded very good predictions!) was that it produced what amounted to a sea-surface spectrum years before an analytic spectral description of sea state was introduced.

In many ways this early work illustrates Walter's modus operandi. Pick a difficult problem with a big payoff, develop physical insight before equations, rely on data and observation, collaborate with colleagues, both scientists and engineers, knit it all together with great mathematical facility, and explain it simply and clearly. When working with colleagues, he always shoulders far more than his share, and even though he does more of the work than anybody, he takes less of the credit. He is particularly insistent that young people, especially students, be recognized for their contributions.

Walter has a self-professed, unique, personal philosophy. He says he chooses not to spend time polishing lectures because students learn more if they participate in halting derivations and have the joy of pointing out blunders. Yet anyone who has heard Walter's clear, concise, organized, and often humorous talks, speeches, and lectures might justly wonder about the truth of this assertion. His colleague Christopher Garrett has accused him of not letting the truth get in the way of a good seminar.

Walter has said, "I do not like to read." It may be true, but he obviously reads voraciously. One of the great joys of his writing is the wonderful way he puts things into historical perspective with his virtually photographic memory of facts and figures that can go back to Devonian times when the year had 400 days! We think Walter's philosophy is like his sense of humor—mischievous, with tongue in cheek.

Walter has modestly written about himself: "During my career I have worked on rather too many topics to have done a thorough job on any one of them; most of my papers have been superseded by subsequent work. But 'definitive papers' are usually written when a subject is no longer interesting. If one wishes to have a maximum impact on the rate of learning, then one needs to stick out one's neck at an earlier time. Surely those who first pose a pertinent problem should be given some credit, not just criticized for having failed to provide a final answer."

The work on waves conducted during the war years was followed by theoretical studies of wind-driven ocean gyres and then on Earth's wobble and spin in the 1950s. Irregularities in the Earth's rotation are an elegant remote sensing tool from which one can infer information about Earth's core, its air and water masses, and global winds. The notion of remote sensing has been fundamental to Walter's science, and it is no surprise that remote sensing is the basis for Walter's major contributions to ocean acoustics—the use of underwater sound to measure physical ocean parameters such as tides, currents, and heat content (Munk, 2006).

Walter spent a decade working on ocean swell (1955-65) and another (1965-75) on ocean tides. His work on swell is reminiscent of his later work on ocean acoustics in that both are of global dimensions, both involved global instrument arrays, and both involved close collaboration with engineers. Aerial photographs taken off the California coast during the war had revealed regular swells that Walter calculated (taking into account the change of direction caused by wave refraction in shallow water) were coming from the south-southwest (SSW) and, incredibly (at that time), must have been the result of winter storms in the southwest Pacific. Furthermore, it appeared that the waves were coming from a source either in or beyond the window between Antarctica and New Zealand, some 5,000 miles away. He and Frank Snodgrass, a talented engineer with whom Walter collaborated for 25 years, installed a global array of six wave stations between New Zealand and Alaska to track the waves. The data confirmed that waves off the California coast can indeed originate in Southern Ocean storms and that some may even start in the Indian Ocean some 10,000 miles distant, halfway around the world.

In the mid-1970s, as a result of serving on the JASON committee (an independent scientific advisory group to the Department of Defense named after a hero in Greek mythology), which was working on anti-submarine warfare problems, it was our luck that Walter was lured into the world of underwater sound. The Navy was concerned about the stability of underwater sound transmissions, which exhibited fluctuations in amplitude and phase that negatively affected the performance of sonar systems. Walter's work on the effects of internal waves on ocean microprocesses naturally led to considering their effect on fluctuations in the speed of sound. It was perfect for Walter—a new problem, an important one, and one that no one had solved.

Characteristically, Walter entered what some had thought was a mature field—after all, ocean acoustics had been studied to death since WWII—turned it upside down, completely upsetting much conventional wisdom (perhaps because he doesn't like to read and wasn't encumbered by prevailing ideas?), and generated an excitement that persists today. Through two major books (Flatté et al., 1979; Munk et al., 1995), numerous publications in the *Journal of the Acoustical Society of America* (JASA) and other journals, invited lectures worldwide, and the education of a generation of students, Walter has fundamentally changed the field by the rigorous introduction of oceanography into underwater acoustics. More than anyone, he has revolutionized our understanding of the intrinsic role oceanography plays in ocean acoustics. His name is associated with the invention of ocean acoustic tomography and ocean acoustic thermometry, with the eponymous Munk sound-speed profile, and with the Garrett and Munk internal wave spectrum.

In 1976, as part of a forerunner experiment to ocean acoustic tomography, one of us (PW), then Walter's graduate student, deployed two acoustic transceivers about 25 km apart and measured the differential travel time of acoustic pulses transmitted in opposite directions, thus providing a measure of the average current between the instruments (Worcester, 1977). In this case, transmission in one direction is aided by the current; in the opposite direction it is opposed by the current. This led Walter, together with Carl Wunsch of MIT, to propose the idea of ocean acoustic tomography

(Figure 2) wherein acoustic transmissions between multiple acoustic sources and receivers, or transceivers, would effectively produce a tomographic cross-section of the enclosed ocean volume (Munk and Wunsch, 1979).

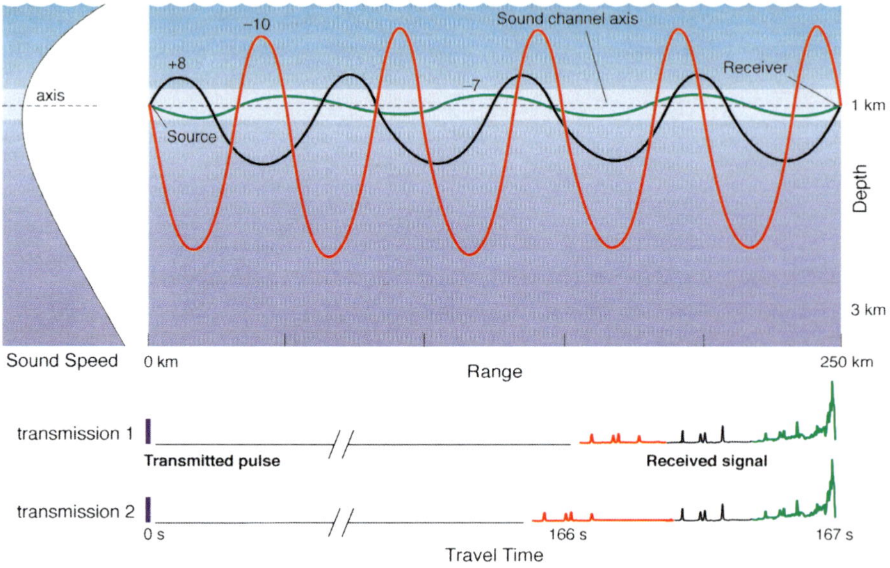

Figure 2. *Cartoon of the Ocean Acoustic Tomography method. The minimum in sound speed provides an axis for an acoustic waveguide. The time sequence of ray arrivals is from steep (red) to moderate (black) to flat (green) launch angles (flat, axial rays cannot generally be resolved). A representative ray for each of the three classes is shown in the upper panel. Transmission 2 (relative to 1) shows earlier arrivals of steep rays but no change otherwise. This is consistent with a warming of the upper ocean (or, less likely, a warming of the abyssal ocean) with intermediary and axial depths unchanged. Modified from Worcester et al. (2005).*

The spatial distribution of sources and receivers provides horizontal resolution, while the up-down cycling of sound paths in the sea provides vertical resolution. The initial idea was to measure the ocean mesoscale, features of roughly 100 km dimension that account for most of the kinetic energy associated with ocean circulation. The mesoscale plays a prominent role in ocean mixing and has unique effects on underwater sound propagation. The first 300 km × 300 km ocean tomography experiment was conducted in 1981 in the Atlantic (Ocean Acoustic Tomography Group, see Behringer et al., 1982). The concept caught on quickly and has resulted in three decades of tomographic experiments by U.S., Japanese, French, and Norwegian scientists, in the Pacific and Arctic Oceans and the Norwegian, Greenland, Mediterranean, Barents, and Philippine Seas.

Figure 3. *Greenland Sea Tomography Experiment. Walter is on the deck of the R/V* Knorr *during the Greenland Sea mooring deployment cruise, September–October 1988. Photo from Peter Worcester*

Never one to settle for 300-km experiments, not with an entire globe beckoning, Walter proposed using ocean acoustic transmissions to measure global ocean warming (Munk and Forbes, 1989). The idea is that sound speed in the ocean increases 4–5 m/s per °C. Thus, for example, taking a typical estimate of 5 m°C/yr warming at the sound channel axis over a 10-megameter (Mm) path, the travel time would change by –0.1 s/yr. Walter convinced the U.S. Navy to assist in a feasibility experiment. In 1991 an array of Navy HLF-4 hydroacoustic transducers was lowered from the R/V Cory Chouest in the vicinity of Heard Island in the Southern Indian Ocean, an almost unique spot that permits ensonification of both the Pacific and Atlantic Oceans. It is also at a high-latitude location where the sound channel axis is shallow, which accommodated the 300 m depth limitation of the HLF-4s.

Transmissions at 57 Hz (to avoid electrical machinery noise at 50 and 60 Hz) with a source level of 221 dB re 1 μPa at 1 m RMS were received at locations throughout both oceans by teams from the U.S., Canada, South Africa, Australia, India, New Zealand, Japan, France, and Russia.

The Heard Island Feasibility Test (HIFT) was followed by the Acoustic Thermometry of Ocean Climate (ATOC) series of experiments during the 1990s in which acoustic sources off the coast of central California and the northern coast of Kauai transmitted to multiple receivers in the North Pacific, producing nearly nine years of time series. A summary of HIFT and ATOC is provided in Worcester et al. (2005). The HIFT was the first underwater acoustics experiment to draw attention to the possible harmful effects of anthropogenic (man-made) underwater sound on marine life, albeit unintentionally. Therefore, Walter has even contributed to biological oceanography as the most prominent instigator of today's vigorous research into the effects of underwater sound on marine mammals.

Walter has said, "I also regret that I am so poor at building and repairing gear (I was sheltered from this as a boy)." Yet he has a deep appreciation for sophisticated measurement techniques, and throughout his career he has partnered with engineers and technicians who were able to build instruments and plan and execute experiments suggested by his theories and speculations. Frank Snodgrass was followed by Walter's graduate student, Peter Worcester, with whom he has worked for over 45 productive years. He also collaborated with Robert Spindel, beginning with the first 3-D ocean acoustic tomography experiment and continuing through the Heard Island and Acoustic Thermometry of Ocean Climate Experiments during the 1980s and 1990s. Walter collaborated with Doug Webb, Woods Hole Oceanographic Institution, on the development of underwater, low-frequency, wideband acoustic sources.

An amusing story comes to mind. We (RS and PW) were at sea. Walter was doing what Walter does at sea, which is to wedge himself into a snug crevice with pad and pencil to work continuously with interruptions only for meals. He was hard at work trying to determine how to instantaneously change the frequency of an acoustic source so as to produce a wideband signal needed for high resolution tomographic measurements. He started with the idea of a vibrating string. What force could be applied to the end of the string to instantaneously stop it from vibrating, thus preparing it for the application of a second force to cause it to vibrate in another mode? There were many equations on lots of paper. People on the ship wanted to know what he was doing. He told them he was trying to figure out how to stop a string from vibrating and demonstrated by tying one end of a string to a bulkhead and moving the other up and down so as to create a standing wave. One of the crew shrugged and said he knew how to do it. He took the string and moved it up and down as Walter had, and then without solving any equations, without putting pencil to paper, he let go of the string. It stopped vibrating, to be sure, but it wasn't the solution that Walter was seeking!

Walter loves going to sea and insists on participating in experiments. Although he's a wonderful shipmate, by his own admission—and our many observations—he is not the most useful fellow on deck. On one expedition Peter Worcester and Walter were conducting an experiment that required dropping SUS charges (small, bomblike explosive devices that are a convenient wideband acoustic source). Spindel's experiment on the same ship called for deploying hydrophone listening devices (sonobuoys), and when he was about to heave one over the side, from up above came the voice of Walter frantically shouting, "Not yet. Not yet, we're not supposed to drop the SUS charges until next Wednesday." He was right about the day but wrong about the device. How could one not know the difference between a bomb and a hydrophone? Ever since, we've been wary about letting Walter near anything other than pencils and paper.

As well known as Walter was, his wife Judith was known nearly as well. They were an extraordinary, inseparable team until her death in 2006, after 53 years of marriage. Wherever Walter went, Judy went, too. Whether abroad or in the U.S., whether for a lecture or a scientific meeting, if Walter was there, so was Judy. She opened their beautiful home in La Jolla to friends, colleagues, and students. A highlight of a Scripps visit was dinner at the Munks', always enriched with good cheer and lively conversation. As Charles Kennel, a former director of Scripps described it, "An evening in her living room was renowned as the quintessential Scripps experience for students, scientists, and legions of friends from around the world." Walter and Judith have two lovely daughters, Edie and Kendall, and three grandchildren. In 2011, Walter married Mary Coakley, and she is carrying on the tradition of La Jolla hospitality. On the occasion of Walter's 65th birthday, Roger Revelle, a former Scripps director and President of the American Association for the Advancement of Science said, "Even if Walter Munk had never done any great scientific work, his life as a husband, father, and friend would still be an inspiration for the rest of us."

For Walter, work is an avocation. At 98 [now 99!] he still rises early every morning excited to tackle the problem de jour. For the last few years he has been working on the generation of low-frequency ocean ambient noise by surface waves. Walter still participates in JASON studies and takes a pro-active interest in the affairs of the U.S. Navy research establishment through his Secretary of the Navy Chair. He has received every conceivable honor, from the National Medal of Science to the Kyoto Prize. He is an Honorary Fellow of the ASA. The United States Navy and The

Oceanography Society established the Walter Munk Award for Distinguished Research in Oceanography Related to Sound and the Sea in his honor (Figure 4).

Figure 4. *The front of the medal for the Walter Munk Award for Distinguished Research in Oceanography Related to Sound and the Sea, granted jointly by the United States Navy and The Oceanography Society. Walter's wife, Judith, designed the medal.*

We and Walter's many other colleagues have been fortunate to have had the opportunity to know and work with him. We take pride in knowing that Walter, a quintessential gentleman, a wonderful story teller, a brilliant scientist, and a wise and generous human being and colleague, who has often claimed to have changed fields every decade, has found acoustics to be so exciting and so compelling that he has stayed with us for more than four!

Biosketches

Robert C. Spindel is the Emeritus Director of the Applied Physics Laboratory of the University of Washington and Professor Emeritus of Electrical Engineering. He received his bachelor's degree from The Cooper Union, New York, in 1965 and his PhD degree from Yale University in 1971. He was a scientist at the Woods Hole Oceanographic Institution from 1972 until 1987 when he joined the University of Washington, where he served as the APL's director until 2003. His research interests are underwater acoustics, acoustic signal processing, and acoustical oceanography. He received the Acoustical Society of America's Silver Medal in Acoustical Oceanography in 2009.

Peter F. Worcester received a BS in engineering physics from the University of Illinois (1968), a MS in physics from Stanford University (1969), and a PhD in oceanography from Scripps Institution of Oceanography, University of California, San Diego (1977). He has been a Research Oceanographer at Scripps since receiving his PhD. His primary research interests are acoustical oceanography and underwater acoustics. Dr. Worcester is the recipient of the 2006 Walter Munk Award for Distinguished Research in Oceanography Related to Sound and the Sea. He is a Fellow of the Acoustical Society of America and the Institute of Electrical and Electronics Engineers.

Credits

Spindel, R. C., and Worcester, P. F. (2016). "Walter H. Munk: Seventy-Five Years of Exploring the Seas," Acoustics Today 12, 36–42.

The original article is available online at http://acousticstoday.org/.

References

Behringer, D., Birdsall, T., Brown, M., Cornuelle, B. et al. (1982). A demonstration of ocean acoustic tomography. Nature 299, 121–125.

Cox, C., and Munk, W. (1954). Measurement of the roughness of the sea surface from photographs of the sun's glitter. *Journal of the Optical Society of America* 44, 838–850.

Flatté, S. M., Dashen, R., Munk, W. H., Watson, K. M., and Zachariasen, F. (1979). *Sound Transmission Through a Fluctuating Ocean*. Cambridge University Press, New York.

Garrett, C., and Munk, W. (1972). Space-time scales of internal waves. *Geophysical Fluid Dynamics* 3, 225–264.

Munk, W. H. (2006). Ocean acoustic tomography: From a stormy start to an uncertain future. In Jochum, M., and Murtugudde, R. (eds), *Physical Oceanography: Developments Since 1950.* Springer-Verlag, New York, pp. 119–138.

Munk, W. H., and Forbes, A. M. G. (1989). Global ocean warming: An acoustic measure? *Journal of Physical Oceanography* 19, 1765–1778.

Munk, W. H., and MacDonald, G. J. F. (1960). *The Rotation of the Earth: A Geophysical Discussion.* Cambridge University Press, New York.

Munk, W. H., Spindel, R. C., Baggeroer, A. B., and Birdsall, T. G. (1994). The Heard Island feasibility test. *Journal of the Acoustical Society of America* 96, 2330–2342.

Munk, W. H., Worcester, P. F., and Wunsch, C. (1995). *Ocean Acoustic Tomography*. Cambridge University Press, New York.

Munk, W. H., and Wunsch, C. (1979). Ocean acoustic tomography: A scheme for large scale monitoring. *Deep Sea Research Part A* 26, 123– 161.

Sverdrup, H., and Munk, W. H. (1947). *Wind, sea, and swell: Theory of relations for forecasting*. U.S. Hydrographic Office Tech. Report No. 1, H.O. Pub. 601.

Worcester, P. F. (1977). Reciprocal acoustic transmission in a mid-ocean environment. *Journal of the Acoustical Society of America* 62, 895–905.

Worcester, P. F., Munk, W. H., and Spindel, R. C. (2005). Acoustic remote sensing of ocean gyres. *Acoustics Today* 1, 11–17.

A Photographic Update 2010-2017

By Deborah Day, Mary Coakley Munk, and Brittany Hook

Seventy Years of Exploration in Oceanography was originally published in 2010. As Walter approaches his centennial, he continues his groundbreaking research and travels the world as an ambassador of science and our oceans. Whether mentoring students at Scripps, participating in global academic conferences, attending the 70th anniversary of D-Day in Normandy, or visiting his beloved homeland of Austria, Walter is a man on the go.

Join us as we cruise through the past eight years...

2010

Walter takes center stage at UC San Diego's 50th Anniversary Founders Day Celebration.
Photo: UC San Diego Publications

8 January Celebration of the 50th Anniversary of UC San Diego.

22 January Commemoration of Walter's gift of his medals to the IGPP Library with thanks to Steve Constable and Jacques Lemire "for housing them so handsomely." According to Walter, the gift is a "symbol of my long career and an expression of my confidence in the survival of our Institute."

4-11 May Receives the Crafoord Prize in Geosciences at the Royal Swedish Academy of Sciences "for his pioneering and fundamental contributions to our understanding of ocean circulation, tides and waves, and their role in the Earth's dynamics." Walter remarked, "The Crafoord Prize is the high point of my seventy years in oceanography…but I've only just begun."

12 May Proposes probing the huge ocean caverns under the Antarctic and Greenland glaciers, thus far inaccessible to observation, with sound waves to study the ocean dynamics at the underside of the floating ice sheets—a daring venture. (Talk given in Lund for the Crafoord Prize.)

14-17 June Notification by Linda Zall, CIA Environmental Specialist and Chair of MEDEA, that, following an eight-year hiatus, MEDEA was to reconvene in Washington, D.C., at the request of the Senate Select Committee on Intelligence. Leon Panetta presided over the meeting.

29 June Proclamation by the San Diego City Council naming Walter Munk Day in the City of San Diego.

17 September Presentation of Österreichische Ehrenkreuz für Wissenschaft und Kunst 1. Klasse (Austrian Cross of Honor for Science and Art, First Class) in Vancouver, British Columbia.

20-21 October Celebration of the 50th Anniversary of the Cecil H. & Ida M. Green Institute of Geophysics & Planetary Physics (IGPP), founded by Walter.

18 November Commemoration of UC San Diego's 50 Fabulous Years at Founders Day Celebration. Walter remembers his close friend and colleague Roger Revelle who played a pivotal role in creating the university.

HM King Carl XVI Gustaf presents the Crafoord Prize to Walter.
©Markus Marcetic, Royal Swedish Academy of Sciences

Admiral Nevin Carr, Commanding Officer of ONR, at the Gala reception.

On Founders Day, Walter reflects on his friendship with Roger Revelle; "No one had a greater influence on my life or my career."
Photo: UC San Diego Publications

Mr. Friedrich Faulhammer, Director General at the Austrian Federal Ministry of Science and Research, awards Walter with the Austrian Medal for Science and Art.
Photo: Michael Sean Lee, Courtesy of OSTA Washington

Mike Dormer's graphic for Walter's Crafoord Prize presentation, *"The Sound of Climate."*

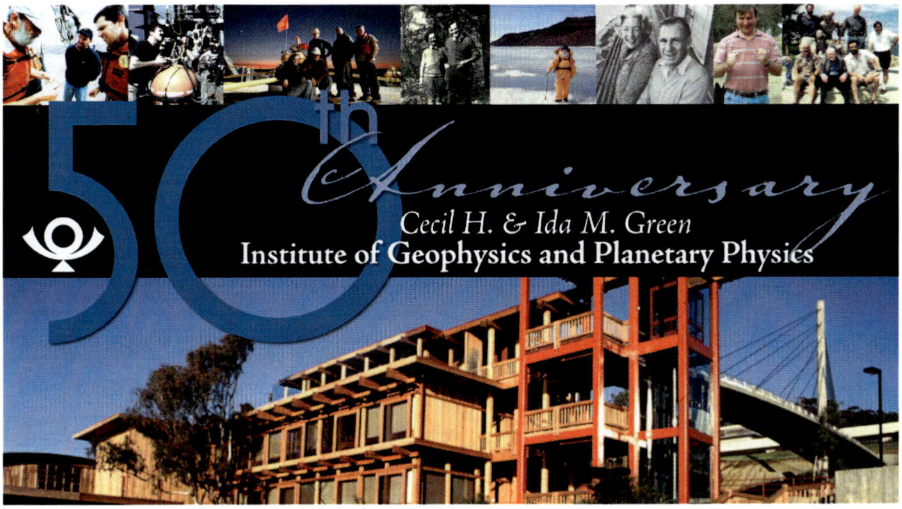

50th Anniversary of IGPP.

2011

Walter, Klaus and Susanna Hasselmann, Bill Farrell, and Ola Johannessen visit a glacier field in Norway.

10-13 March Birch Aquarium Whale Watching trip to Scammon's Lagoon in Baja with Nigella Hillgarth, Executive Director. A great time was had by all!

10 May Proclamation by Supervisor Pam Slater-Price naming Walter Munk Day in the County of San Diego.

5-10 June Presentation of "Monitoring Under-Ice Cavities with Sound" at the Scripps International Symposium on Interactions of Ice Sheets and Glaciers by Peter Worcester and Walter.

12 June Walter makes an impassioned plea to save the iconic Scripps Library at a faculty meeting and ends up in the hospital with ventricular tachycardia. In typical Munk fashion, Walter proposes an impromptu wedding. The ceremony between Walter and Mary takes place in the Bob and Betty Beyster Garden at Sulpezio Hospital with Dr. Kirk Peterson as Walter's Best Man.

18 June Selection of Walter as one of the "50 Leaders" for the UC San Diego Alumni Celebration "50 Years/50 Leaders."

12-16 September Participates in Venice Conference joined by Charlie Kennel, Ken Melville, Giovanni Cecconi, and numerous other experts to review progress of the MOSE flood gates and lay plans to mitigate Venice's vulnerability due to the effects of climate change and sea-level rise.

7-15 November Presentation of "Waves Long/Short" by Bill Farrell and Walter in honor of Klaus Hasselmann's 80th birthday celebrated in Hamburg, Germany, followed by a trip to Klaus and Susanna's home in Glückstadt.

15-18 November 25th anniversary celebration of the Nansen Institute in Bergen, Norway, honoring Ola Johannessen, its founder and director for many years.

19-28 November Visiting the Wunsch's at Oxford University, where Carl is on sabbatical as the George Eastman Visiting Professorship in the Department of Atmospheres, Oceans and Planetary Physics, Fellow at Balliol College.

Bob and Ann Dynes, Gustaf Arrhenius, Terry Kraszewski, Audrey Evans, Mary and Walter, and Nigella Hillgarth enjoying a Mexican feast following a great day of whale watching.

Scripps Oceanography alumni Patricio Bernal, Megan MacArthur, Walter, and Marcia McNutt were honored at UC San Diego's "50 Years/50 Leaders" event. Photo: UC San Diego Publications

Walter and Mary's wedding cake.

Carl Wunsch and Walter looking very dapper in High Table attire!

2012

Dr. Asahiko Taira, Director of JAMSTEC, with Holly Given, Mary and Walter, and members of his staff dressed in their ryokan for dinner following their visit to D/V *Chikyu.*

26 January Presentation by Walter at the La Jolla Riford Center's Distinguished Speaker Series.

7 April La Jolla Historical Society talk titled "Science and Technology on the Homefront."

28 April Participates in the Roast of Monmouth University President Paul Gaffney II, retired Vice Admiral in the U.S. Navy.

11 May Invited by Tom Olson to speak at the Norwegian Fish Club of San Diego; membership includes Norwegians, Norwegian-Americans, and retired naval officers and educators interested in Norway and its ties to the U.S.

15-21 June Presentation of "Wind Drag" paper with Bill Farrell at the 29th IUGG Conference on Mathematical Geophysics at the National Museum of Scotland, Edinburgh.

22-28 June Visits Jim and Annamaria Cairns in Urbino, Italy.

29 June-1 July Attends "Oil Spills in Sea Ice" conference hosted by Pia and Peter Wadhams in Fermo, Italy.

16-18 August Attends keel-laying ceremony for R/V *Sally Ride* at Dakota Creek Industries Inc. shipyard in Anacortes, Washington.

5-7 September Delivers talk at UC San Diego "Sea Level Rise Projections for 2100-2300; Remarks on Sea Level Rise" at the Association of Pacific Rim Universities (APRU) Sustainability and Climate Change Program International Workshop on Coastal Cities, Climate Change and Sea Level Rise.

3 October Presentation of "Acoustic Probing of the Ocean Wedge Under an Ice Sheet" with Peter Worcester and Matthew Dzieciuch at AMISOR Workshop.

8 October Presents "Where the Swell Begins" for the Fall Lecture Series at Birch Aquarium; discusses how World War II and measurements of Guadalupe Island led to this discovery and what it means for surfers today.

16-31 October Visit to CSIRO in Hobart, Tasmania, Australia for the Southern Ocean Observing System (SOOS) meeting. Walter presents a paper on the use of acoustic tomography to probe the ocean wedge between the floating Antarctic ice sheets and the sea floor, where man has never been. Side trips include the Mona - Museum of Old and New Art, Andrew and Caroline's "shack," Mt. Wellington, and nightly happy hours at the ART MOB.

1-4 November Begins three-day adventure in Tokyo with Professor Katsu Michibayashi, Walter and Mary's guide through the history and culture of Japan, ending at a small fishing village on the Kii Peninsula where they take a helicopter to D/V *Chikyu.*

5 November Visit to D/V *Chikyu*, capable of drilling deeper at sea than any other science drilling vessel to date; research program involves 60 scientists from 16 nations.

7 November Panel discussion at the University of Tokyo followed by reception honoring Walter.

10-14 November Honored guest at Kyoto Prize Ceremony. Walter's first trip back since receiving the 1999 Kyoto Prize.

Walter lands on the D/V *Chikyu*, 80 kilometers out to sea off the Kii Peninsula.

JAMSTEC President Asahiko Taira informs Walter that the library aboard the D/V *Chikyu* will be named for him, a great honor.

Walter enjoys a visit with Her Imperial Highness Princess Takamado at the Kyoto Prize Ceremony.

Commemorative gift from Walter to Paul Gaffney; Mike Dormer's cartoon recalls Admiral Gaffney's trip to Japan to celebrate Walter's receipt of the Kyoto Prize in 1999.

Walter and Mary with Phil McGillivary, Peter Wadhams, and Navy Lt. William Eucker taking a well-deserved break in the best gelateria in Fermo, Italy.

Matt Dzieciuch and colleagues join Walter for an International Conference in Tasmania where Walter also visits long-time friends, the Churches and Forbes.

A rare opportunity to spend time with the Inamores at the Kyoto Prize Ceremony.

Walter, Nigella Hillgarth (Marine Marshall), and UC San Diego Chancellor Pradeep Khosla (Innovation Marshall) join Santa and his sleigh for the La Jolla Christmas Parade.

2013

Title graphic for Walter's lecture at the American Philosophical Society meeting.

9 January Delivers talk "Where the Swell Begins" at CASPO meeting at Scripps.

5 February Gives an invited talk at Penn State.

11 February Final visit to the Whaling Bar (1949-2013) at the La Valencia Hotel, a place that holds many special memories.

16 February Surfline honors Walter Munk as "Father of Surf Forecasting" at the Groundswell Society 11th Annual Surfing Arts meeting.

4-7 April Visits Jim and Daphne Jameson at La Mancha in El Barril, Baja California, Mexico, with a side trip to San Ignacio Lagoon to see the whales.

25 April Delivers the distinguished Penrose Memorial Lecture titled "The Perfect Storm" at the Annual American Philosophical Society (APS) meeting in Philadelphia, followed by an award presentation from Surfline's Mark Willis (Director of Forecasting, Chief Meteorologist) and Graeme Rae (Director of Research and Development, Chief Scientist) for Walter's contributions to surf prediction for soldiers and surfers alike.

27-28 April Gala celebration commemorating JASON'S 50 years of service to our Nation; Walter has been a member since 1961.

29 April Walter attends President Obama's speech for the 150th anniversary of the National Academy of Sciences in Washington, D.C.

14-16 May Alfredo Giron, student, invites Walter to give a talk at CISESE, Ensenada, Baja California.

15 June Receives Aquarium of the Pacific's Ocean Conservation Award from Jerry Schubel, President and CEO.

20 June Attends dedication of Ellen Browning Scripps Memorial Pier.

10 July Celebrates the closure/renovation of the NOAA Fisheries buildings.

19 July Participates in Ocean Exploration 2020: A National Forum at the Aquarium of the Pacific.

26 July Presents "The Perfect Storm" about the planning for D-Day landings at "Dining Out," a formal event at the Naval Oceanography Special Warfare Center in San Diego that provides oceanographic and meteorological support to the U.S. Navy.

15 -19 September Visit with Arthur and Carol Baggaroer at Woods Hole and WHOI for meeting of the Navy Chairs.

20-22 September Visits Maurizio Seracini, who discovered the lost Leonardo da Vinci painting behind Giorgio Vasari's mural "The Battle of Marciana."

23-25 September Honored guest at 2013 International Conference on Marine Data and Information Systems in Lucca, Italy; named Pioneer of Physical Oceanography by the Historical Oceanography Society.

25 September Vacations with Jim and Annamaria Cairns in Liguria; visits Carrara marble mines, and dines at Al Cavatore, Walter's favorite restaurant.

12 October Attends the Austrian Research and Innovation Week held at the Annenberg Community Beach House in Santa Monica. Walter meets Walter Kohn and Hannes Androsch, current owner of the Austrian Salt Mines where Walter's step-father was once Director General.

15-16 November Receives Roger Revelle medal at UC San Diego's Founders Day celebration. The Revelle Medal, revived by Chancellor Pradeep Khosla, recognizes current and former UC San Diego faculty members for sustained, distinguished, and extraordinary service to the campus. Arthur and Molli Wagner and Conrad Prebys were the recipients of the 2013 Chancellor's Medal.

A fond farewell to the iconic Whaling Bar at La Valencia.

Ken and Sabina Wallach, Daphne and Jim Jameson, Walter and Mary head to El Barril, Mexico for a well-deserved break!

Surfline's Mark Willis and Graeme Rae present award to Walter at APS meeting.

A member going on 56 years.

Alfredo Giron invited Walter to speak to his class at CISESE.
He is currently Octavio Aburto's PhD student at Scripps.

Jerry Schubel, Walter, Sylvia Earle, John Molina, and Don Walsh at Aquarium of the Pacific's award ceremony.

Cathy and Steve Constable with Walter at Scripps Pier dedication.

Walter "Dining Out" with the Navy SPAWAR.

Walter at the Historical Oceanography Society meeting with
Anna and Federico De Strobel.

Walter with Hannes Androsch
in Santa Monica.

Walter and Mary attend UC San
Diego Founders Day 2013.
Photo: Erik Jepsen/
UC San Diego Publications

Medalists Walter, Arthur and Molli Wagner, and Conrad Prebys
with Chancellor Khosla.

Walter receives the Roger Revelle medal from Chancellor Pradeep Khosla. Photo: Erik Jepsen/ UC San Diego Publications

2014

Pope Francis meets with conferees, including Walter, in the Vatican parking lot.
Photo: Andy Revkin/The New York Times

10 February Meets with the Mexican Consulate in San Diego to gain support for upcoming expedition to Cabo Pulmo. Lunch with Cónsul General Remedios Gómez Arnau.

10 February Presentation, by Nigella Hillgarth, of "the gift of adoption of the Giant sea bass, the longest living fish at the Birch Aquarium at Scripps that is expected to live another 50 years!"

2-9 March Visit to Cabo Pulmo, Baja, Mexico, with Giuseppe Notarbartolo di Sciara, Eliana Álvarez Martínez, Andrea Swift, Octavio Aburto, Josh Stewart, Tim Rowell, Brad Erisman, Nigella Hillgarth, Rick and Patty Elkus, Larry and Mary Jameson, and the Castro Family so Walter could see his namesake, *Mobula munkiana* (Munk's Devil Ray) for the first time. While filming "Spirit of Discovery," Scripps scientists studied their behavior. Why do they jump? Walter says, "It's for the joy of it!"

10-15 March Receives Explorers Club Medal "for his contributions to scientific exploration and discovery." Walter and Scripps researcher Octavio Aburto return Flag #221 to President Alan Nichols, with VP of Flag & Honors Constance Difede and Club Medalist Dr. Sylvia Earle.

2-6 May Invited by Scripps scientist V. Ramanathan to participate in the Vatican Symposium on climate change, a first ever joint workshop of the Pontifical

Academy of Sciences and the Pontifical Academy of Social Sciences, entitled "Sustainable Humanity, Sustainable Nature: Our Responsibility." Pope Francis was inspired to include climate change and its impact on the poorest three billion in his Encyclical Letter, Laudato Si'.

16-31 May Gives a talk at the Nansen Environmental and Remote Sensing Center in Bergen, Norway, celebrates Norwegian Constitution Day, and visits his mentor's daughter Anna Sverdrup Hamre.

3 June Receives Medal of Honor of the French Navy, presented for his work on wave predictions during World War II, followed by a celebration at La Tour d'Argent.

6 June Invitations from United States President Barack Obama and French President François Hollande to attend the 70th Anniversary of D-Day in Normandy.

13 June Participates with Octavio Aburto in a press conference led by Institute of the Americas at UC San Diego, with journalists from many Latin American countries in attendance. They discuss the recent Explorers Club expedition to Cabo Pulmo to study the behavior of the *Mobula munkiana* (Munk's Devil Ray).

14 June Receives The Ralph B. White Memorial Award in recognition for unique contributions to Oceanographic Exploration and the Conservation of the Seas from the Southern California Chapter of the Explorers Club.

28 June Attends world premiere of *Mexican Seas / Mares Mexicanos* exhibit at Birch Aquarium at Scripps. Photographs by Scripps scientist Octavio Aburto "capture the art of science and seek to inspire ocean stewardship." A photograph of Munk's Devil Ray is included in the collection.

31 July "The Gift of Seiche." Ceremony held at Seiche to gift Walter's home to Scripps Institution of Oceanography to serve as the home of its director in the future. A copy of the original key was presented to Chancellor Pradeep Khosla, UC Regent Sheldon Engelhorn, and Scripps Director Margaret Leinen for safe keeping. Walter said that he and Judy always thought that the Scripps director should "live above the shop."

9 August Travels with Margaret Leinen to Anacortes, Washington for the christening ceremony of R/V *Sally Ride*.

28 August Celebrates CharlieFest symposium in honor of former Scripps Director Charlie Kennel's 75th birthday.

3-15 September Invitation from Frans-Peter Lam to ESOMM International Meeting in Amsterdam where Walter delivered a talk, titled "Acoustic Monitoring of Global Ocean Warming: The Very First Encounters with Concern About Marine Mammals and Sound, A Brief Personal Historical Narrative." Resulting article was later published in *Aquatic Mammals*. Walter then went to Holland where he spoke at the Royal Netherlands Institute for Sea Research and met with students.

12 October Attends Annenberg Retreat at Sunnylands, co-hosted by H.S.H. Prince Albert II of Monaco Foundation and Scripps Institution of Oceanography at UC San Diego, to address rising sea levels and ocean acidification.

22 October Nancy Knowlton, Sant Chair for Marine Science, gives John Orcutt, Peter Mikhalevsky, and Walter a private tour following a special meeting at the Smithsonian.

24 October Anna, Harold Sverdrup's daughter, and Walter enjoy the sculpture created by Niki de Saint Phalle outside the Mengi Museum in Balboa Park.

Meeting with the Mexican Consulate: Antonio Diaz, Octavio Aburto, Gerardo Toledo, Mary and Walter, Remedios Gómez Arnau, Exequiel Ezcurra, Jesus Lopez, and Gabriela Pinzon.

Dedication ceremony: School children honor Walter and Giuseppe with replicas of Walter's namesake, mounted along the entry to Cabo Pulmo.

"Swimming in the Air," Munk's Devil Ray in Cabo Pulmo, Baja, Mexico. Photo: Octavio Aburto / iLCP

David Castro and Walter getting a different view of Cabo Pulmo.

Giuseppe Notarbartolo di Sciara and Walter enjoy the great outdoors in Cabo Pulmo.

The remarkable Castro Family with Walter and other Scripps researchers in Cabo Pulmo. They are displaying the Explorers Club Flag that was later returned at the Annual Meeting. Photo: Patty Elkus

Walter is honored with the Explorers Club's highest award, "The Explorers Club Medal" and is shown here returning Flag #221 from the expedition to Cabo Pulmo. During the award ceremony, he received a two-minute standing ovation by 900 people in the packed Waldorf Astoria Ballroom. Photo: ECAD

Walter participates in the Vatican meeting. Photo: Matthew Siegfried

Scripps graduate student Matthew Siegfried, Walter and Mary, Scripps graduate student Emily Kelly, and Scripps alumnus Danny Richter attend the Vatican conference. Photo: Matthew Siegfried

Walter, friends Peter and Pia Wadhams, and the Scripps team in the dining room of Domus Sanctae Marthae where Pope Francis shares meals with Vatican guests.

Walter with a member of the Swiss Guard.

Anna Sverdrup Hamre and Walter with film crew.

Examining the wine list at Tour d'Argent.

Walter and General Frachon, Director of the Service Hydrographique et Océanographique de la Marine (SHOM) at the Hotel de la Marine, Paris.

Under a rare Normandy sun, Walter enjoys breakfast at Damien Leloup's mother's home in Port-Guillaume, only meters away from the Normandy beach landing where the bombarding ships "BB Roberts," "BM Mauritius," and "CL Arethusa" anchored some 70 years earlier.

Mary and Walter join Secretary of State John Kerry at the 70th Anniversary of D-Day in Normandy.

Walter at the D-Day Dedication of the Caen Memorial Museum at Pointe du Hoc.

Walter with D-Day Veterans in Normandy.

In Flanders fields the poppies grow
Between the crosses, row on row,
That mark our place; and in the sky
The larks, still bravely singing, fly
Scarce heard amid the guns below.

We are the Dead. Short days ago
We lived, felt dawn, saw sunset glow,
Loved and were loved, and now we lie
In Flanders fields.

Take up our quarrel with the foe:
To you from failing hands we throw
The torch; be yours to hold it high.
If ye break faith with us who die
We shall not sleep, though poppies
grow
In Flanders fields.

Lieutenant Colonel John McCrae
WWI soldier, physician and poet

Walter reflects on D-Day. Photos: Damien Leloup

UC San Diego Chancellor Pradeep Khosla; Tam O'Shaughnessy, cofounder and CEO of Sally Ride Science; Scripps Director Margaret Leinen; and Walter attend the christening of R/V *Sally Ride* in Anacortes, Washington.
Photos: Jeff Dillon/Scripps Communications

Walter and Rear Admiral Matthew Klunder, Chief of Naval Research, in front of R/V *Sally Ride*.

Walter at the christening ceremony for R/V *Sally Ride*.

Walter takes a tour of R/V *Sally Ride*.
Photo: Robert Monroe/Scripps Communications

"The Gift of Seiche," Walter's home, is endowed to Scripps.
Photos: Damien Leloup

Octavio Aburto and Walter stand next to Walter's namesake at the opening festivities for the world premiere exhibit of *Mexican Seas / Mares Mexicanos* at Birch Aquarium at Scripps.

Anna, Harald Sverdrup's daughter, and Walter enjoy sculpture by Niki de Saint Phalle in Balboa Park.

2015

Double Trouble!
Photo: Tenzin Choejor courtesy of the Office of His Holiness the Dalai Lama

9-13 January Snorkeling with the endangered Manatees in Crystal River, Florida, at the Three Sisters Springs.

13 March Celebrating the Revelle College 50th birthday with a tribute to its namesake and UC San Diego Founder Roger Revelle, his work and legacy, and the way his work has influenced the areas of sustainability then, over the years, and into the future.

14 April Celebration of the 40th Anniversary of the National Academy of Sciences.

16 April ARCS Scientist of the Year Honorary Chair.

5 June Revelle College 50th Alumni Celebration at Seiche; ceremonies were held for the first two graduating classes from 1965 and 1966.

13-22 June An unforgettable trip down memory lane with Rick and Patty Elkus including tea at Walter's boyhood home in Vienna that now serves as the South Korean Embassy; a visit with Professor Anton Zellinger at the Austrian Academy of Sciences; a quick flight to Venice with Francesca von Habsburg to meet Giovanni Cecconi and check on the status of the "Gates"; and dinner at his vacation home in Altaussee with the von Fuersters, current owners.

22 June-2 July Presents talks on "Internal Waves" and "Acoustic Monitoring of Global Ocean Warming" in Prague, Czech Republic at IAPSO symposia of the 26th IUGG General Assembly.

5-7 July Participates in Global Compassion Summit at the University of California Irvine, where Walter gives talk on climate change and meets with His Holiness the Dalai Lama.

31 July Invitation from V. Ramanathan to serve on the UC Carbon Neutrality Summit Committee with initial meeting at UC Irvine. Walter compared the current effort by UC with the time when all the campuses came together during WWII to work on Anti-Submarine Warfare (ASW) methods for locating German submarines that were sinking Allied shipping at will. Within three and a half years, by working together, the Allies were successful in their landing at Normandy.

6 August Photo of five generations of ocean scientists at Scripps. Walter, who received his doctorate from Scripps in 1947; Munk's student, oceanographer Charles "Chip" Cox, who received his PhD from Scripps in 1954; Cox's student, oceanographer Michael Gregg, who received his PhD at Scripps in 1971; Gregg's student at University of Washington, oceanographer Jennifer MacKinnon, who received her PhD in 2002; and Caitlin Whalen, a student of MacKinnon and oceanographer Lynne Talley, who received her PhD in 2015.

20 August *The New York Times* proclaims Walter Munk "Einstein of the Oceans."

15 September Approval of Scripps Center for Marine Archaeology at UC San Diego, co-founded by Walter and Damien Leloup.

18 September Visit with Walter's namesake, *Mobula munkiana*, at Scripps Marine Vertebrates Collection.

26 September In New York for events related to Pope's visit to U.N.

19 October Special present from Scripps scientist Greg Rouse to celebrate Walter's 98th birthday; Greg named a new species of deep-sea worm (Hesionidae, Annelida) *Sirsoe munki* in Walter's honor.

27 October Launch of the University of California Carbon Neutrality Initiative; pledge to become carbon neutral by 2025.

3 November Recipient of the American Polar Society Medal.

9-12 December Special Guest at opening exhibit of "Road to Tokyo" at the National World War II Museum, New Orleans.

Scripps Center for Marine Archaeology logo

Walter snorkeling with the manatees. Photos: Damien Leloup

Charles Perrin, Walter, Revelle College Provost Paul Yu, Tom Bond, Sarah Gille at Revelle College 50th anniversary event. Photo: Erik Jepsen/UC San Diego Publications

Celebrating Revelle College's 50th Class Reunion at Seiche, Walter's home, where their graduation was held.

Pondering sliding down the banister.

Walter's home in Vienna, now the South Korean Embassy.

The tree Walter climbed as a child.

Professor Anton Zeilinger, a quantum teleportation expert, and Walter at the Hedy Lamarr Quantum Communication Telescope in Vienna. Lamarr, a successful scientist and film star, was home-schooled with Walter.

Touring Prague with Mary and Patty Elkus in a 1929 Praga Alfa! Photo: Patty Elkus

In celebration of the Dalai Lama's 80th birthday, V. Ramanathan and Walter presented him with a framed image of a deep-sea worm named *Sirsoe dalailamai* in his honor. The worm, discovered by Scripps researcher Greg Rouse, lives in methane seeps off Costa Rica. *Sirsoe dalailamai* was named after the spiritual leader in part because it eats methane, giving back to the environment, just as His Holiness has given back to humanity.

Giovanni Cecconi, Walter, and Francesca von Habsburg visit the Venice Gates, a project Walter worked on in the early 1970s with Roberto Frassetto.

Walter celebrates the 40th Anniversary of the National Academy of Science with Ralph Cicerone (President) and his wife Carol, a grand affair!

Five Generations of Ocean Scientists: Walter, Chip Cox, Michael Gregg, Jennifer MacKinnon, and Caitlin Whalen. Photo: Jeff Dillon/Scripps Communications

Walter visits the Scripps Marine Vertebrates Collection with graduate students and postdocs.

Collection Manager H.J. Walker is holding a familiar specimen: Munk's Devil Ray!

A painting of Eggelgut, Walter's childhood home in Altaussee, Austria, gifted to Walter on his 98th birthday by well-known Altaussee artist Horst Jandl.

Ola and Bente Johannessen from the Nansen Center in Bergen, Norway visit Walter at his home in La Jolla.

Walter and Jesse Ausubel in Manhattan.

Scripps Professor and Curator Greg Rouse discovers a new species of deep-sea worm and names it *Sirsoe munki* to honor Walter.

Walter surrounded by the Victory Belles at the opening exhibit of "Road to Tokyo" at the National World War II Museum, New Orleans.

2016

Speaking to a full house on "An Oceanographic Perspective" discussing the evolution of his research.

31 January Interviewed by Max Guinn, a 15-year-old entrepreneur and founder of Kids Eco Club (www.kidsecoclub.org), an organization of over 100,000 K-12 students, which establishes and supports school based environmental clubs. Guinn's school assignment turned into a documentary film and *Huffington Post* article.

7-9 February Travels to Rochester, Minnesota, with V. Ramanathan to meet with His Holiness the Dalai Lama, Lama Tenzin Dhonden, and Ann Curry.

21-26 February Presentation of the Munk Medal granted jointly by the Oceanography Society, the Office of Naval Research, and the Office of the Oceanographer of the Navy to Carl Wunsch at The Ocean Sciences Conference (TOS) in New Orleans. Scripps oceanographer Lynne Talley received the Albatross Award; Walter was the second ever recipient in 1959.

29 February-1 March Walter and Mary escape to the Bahamas for a short visit with Katrina Vanderlip, artist and environmentalist.

12 March Attends annual Explorers Club Gala in New York City.

15 April Attends ribbon-cutting ceremony at the Nimitz Marine Facility (MarFac), home port and technical support center for Scripps' world-class oceanographic research fleet in Point Loma, to celebrate the completion of a five-year, $20.9 million rehabilitation project.

24-27 April Guest Lecturer at the University of Delaware (UD) College of Earth, Ocean, and Environment, invited by Mohsen Badley and Fabrice Vernon. UD Acting President Nancy Targett introduces Walter, stating, "His lifetime of pioneering research, his passion for teaching and his many accolades make Dr. Munk something of a legend in this field. It is no exaggeration to say that he built and shaped the way we study and understand the ocean."

28-30 April Travels to Philadelphia to attend annual meeting of the American Philosophical Society, Walter's favorite society.

1 May Annual AWFUL WAFFLE breakfast with Paul Chamberlain and Scripps postdocs at Surfside.

6 May Walter and Damien Leloup meet with Debra Brice's Science Crusaders from the San Marcos Middle School at Birch Aquarium. Students proudly display an ROV they built.

10 May Receives honor as a member of the inaugural class of Revelle College Faculty Fellows.

14 May Officiates at the marriage of his good friends Tiffany Beres and Damien Leloup at Seiche.

26 May "What the Sea Gives Me," directed by Pierce Kavanagh featuring Walter and other ocean advocates, presented by the San Diego Surf Film Festival.

4 June 38th Annual Alumni Celebration honoring Rear Admiral Timothy Gallaudet, USN Oceanographer and Navigator of the Navy, for his research at Scripps Oceanography and his role in overseeing Navy exploration of the sea and sky.

12 August PhD awarded to Chris Verlinden, Walter's latest doctoral student.

20 August Salk Institute Annual Gala where Leon Fuerth, Walter Munk, and Al Gore enjoy a visit.

4-6 October Celebrates Ken Melville's 70th birthday and oceanographic career.

20 October Meeting for climate change committee members led by V. Ramanathan and Mario Molina. Participants, including Walter, author report "Well Under 2 Degrees Celsius: Fast Action Politics to Protect People and the Planet from Extreme Climate Change." Report was published and delivered by Ramanathan to the U.N. on 18 September 2017.

6 November Annual Scripps professors' children's holiday party at Seiche.

25 November-3 December Invitation by Bruce Howe to speak at the Acoustical Society of America's (ASA) 25th Annual Meeting in Honolulu. Visits Ford Island,

USS Missouri, Pacific Aviation Museum, and NOAA Inouye Pacific Regional Center.

4 December Informed by Don Walsh that Walter has been appointed a member of the Ocean Elders…Walter is the eldest!

Walter with Max Guinn visit following "an excellent interview."

Walter congratulates Carl Wunsch on being awarded the Munk Medal.

Celebrating Carl Wunsch's selection as recipient of the Munk Medal. Peter Worcester, Carl, Walter and Mary, Marjorie Wunsch, John Colosi.

Scripps oceanographer Lynne Talley has the honor of receiving the Albatross Award from the American Miscellaneous Society.

Walter with UD undergraduate students taking an "Ocean Science" course.

Another first for Walter…Minister for a Day!

San Marcos Middle Schoolers meet with Walter and Damien Leloup at the Birch Aquarium at Scripps.

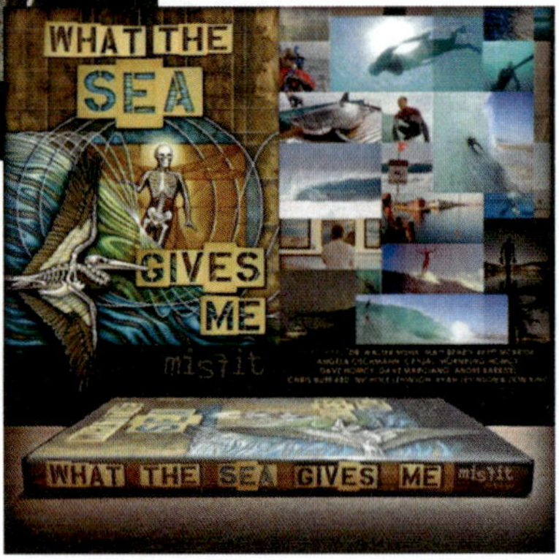

Pierce Kavanagh debuts movie "What the Sea Gives to Me" featuring Walter and other ocean advocates.
Photos: Pierce Kavanagh

Walter attends the MarFac reopening ceremony with Scripps friends: Executive Director of Development Stuart Krantz, Director's Council Member Julie Brown, Emeritus Director's Council Member Megan Bailiff, Associate Director Bruce Appelgate, and Director Margaret Leinen.

Celebrating Walter's 98th birthday with Jim and Anne Hubbell, Rick and Patty Elkus in Julian at the Drug Store…best fried chicken and strawberry sodas in California! Photo: Patty Elkus

A visit from Captain Paul Spear, U.S. Navy, OPNAV, who has trouble making it to La Jolla without a detour to his favorite bakery.

Rear Admiral Tim Gallaudet honored as "Distinguished Alumnus" at UC San Diego's Alumni Gala.

Walter with his most recent PhD student, Dr. Chris Verlinden, and Jacquelyn Kubicko, a Scripps summer intern and advisee.

Honolulu ASA meeting concludes with a luau.

Walter with his student Bruce Howe, Scripps PhD, 1986.

Walter with Ann Curry at the Explorers Club Gala in NYC.

Al Gore visits with Walter and Leon Fuerth, his national security adviser.

Linda Zall, CIA Environmental Specialist, spearheaded the film project to tell the story of MEDEA. Walter and Mary visit Linda and Lens in Orlando to celebrate the newlyweds.

2017

Ocean Elders Meeting in Monaco with H.S.H. Prince Albert II.

11 February Opening of traveling exhibit "Weather on Steroids: the Art of Climate Change Science" at the La Jolla Historical Society. Walter worked with Curator Tatiana Sizonenko and Science Consultant Alexander (Sasha) Gershunov, a Scripps scientist, to produce his contribution, an essay titled, "Climate Change: A Look at 80 Years of Work at Scripps Institution of Oceanography."

14-16 February Oceanology International North America (OINA) "Catch the Next Wave" conference honors Walter "for his outstanding contribution to ocean science and technology."

28 February First screening on Swiss national TV of "The Warning: The Secret American & Russian Files on Climate Change," a documentary about MEDEA written and directed by Paul Jenkins. "The Warning" tells the unknown story of how the CIA engaged in the controversial area of climate change and extreme weather events in the aftermath of the Cold War and partnered with its erstwhile adversary, the GRU, Russia's main military intelligence organization.

23-29 March Explorers Club Gala on Ellis Island followed by a holiday in Antibes.

30 March Walter joins the Comité d'Honneur of the Institut Océanographique Paul Ricard.

31 March-4 April Monaco Ocean Week meeting with H.S.H. Prince Albert II.

6 April Club Altura Women's Luncheon talk followed by lively discussion about Walter's career and climate change.

10-12 May Celebrates 90th Anniversary of California Institute of Technology Geological and Planetary Sciences with Mel Simon and Holly Given. Walter was one of two soon-to-be centenarians to give a talk.

15-17 May Munk Centennial Symposium "International Wave Breaking, Turbulence, and Ocean Mixing" organized by Scripps scientists Rob Pinkel and Matthew Alford.

28 May Walter honored with Doctor of Humane Letters by the University of San Diego.

1 June Launch of Scripps Center for Marine Archaeology at UC San Diego, made possible through generous donations from Ellen Lehman and Charlie Kennel, Becky and Charlotte Rygh, Bill and Kathy Scripps, James and Daphne Jameson, and Mary Jameson.

7 June Walter returns to his student residence at Columbia's International House in Manhattan.

4-8 June Walter addresses U.N. Meeting in New York for "Ocean Week" where "The First Global Integrated Marine Assessment; World Ocean Assessment 1" is submitted, the culmination of the work of nearly 600 scientists and experts "examining the current state of knowledge of the world's oceans and the ways in which humans benefit from and affect them."

9 June Chosen as Honorary Chair for Mainly Mozart; Gala hosted at Seiche.

16-17 June Luncheon and private meeting with the Dalai Lama following His Holiness' address to UC San Diego's Class of 2017.

4 July Celebrates more than 30 years of hosting 4th of July barbeques for JASON families at Seiche.

25 July–4 August Returns to his childhood homes in Vienna and Altaussee, followed by trips to Tuscany for the Andrea Bocelli Concert and Lake Como to spend time with Jim and Annamaria Cairns.

29-30 August Munk Centennial Symposium "Ocean Acoustics, National Security, and Walter Munk" organized by Scripps scientists Peter Worcester and Bruce Cornuelle.

16 September Celebration of Birch Aquarium's "25 Years on the Hill."

18 September Report on Extreme Climate Change presented by V. Ramanathan at the United Nations.

7 October Attends film debut of "Spirit of Discovery" at the San Diego Film Festival.

18 October Celebration of Walter Munk Day in the City of San Diego. Ribbon-cutting ceremony dedicating Walter Munk Way at the La Jolla Shores Boardwalk in Kellogg Park followed by a reception at Scripps.

19 October Celebration of Walter's 100th birthday at Birch Aquarium at Scripps followed by a party at IGPP.

26 October Dialogue with H.S.H. Prince Albert II moderated by Scripps Director Margaret Leinen.

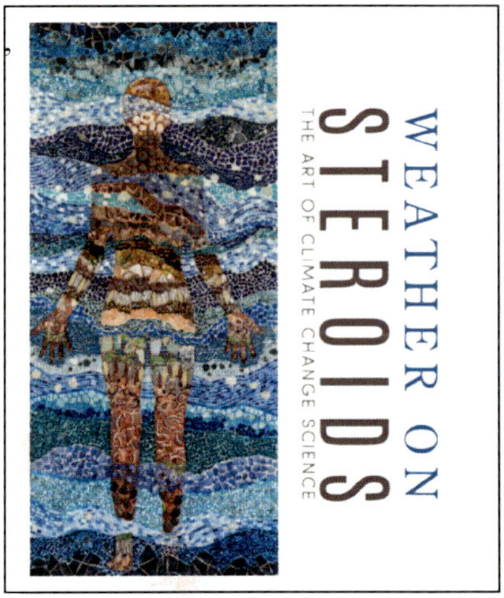

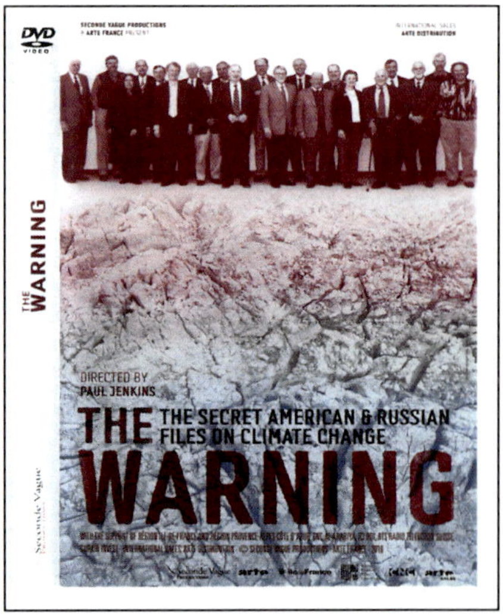

Walter was involved in two creative projects in 2017: "Weather on Steroids," an art exhibition, and "The Warning," a film documentary.

Walter with
President Patricia Ricard,
granddaughter of the founder
of the Institut Océanographique,
and Professor Nardo Vicente,
Île des Embiez, France.

H.S.H. Prince Albert II and his distinguished colleagues celebrate Monaco Ocean Week.

Walter celebrates the 90th Anniversary of the Division of Geological and Planetary Sciences (GPS) at Caltech with Mel Simon and Holly Given.

"Internal Wave Breaking, Turbulence, and Ocean Mixing"
Munk Centennial Symposium at Scripps, 15-17 May 2017.
Photo: Jeff Dillon/Scripps Communications

University of San Diego President James Harris confers Walter Doctor of Humane Letters.

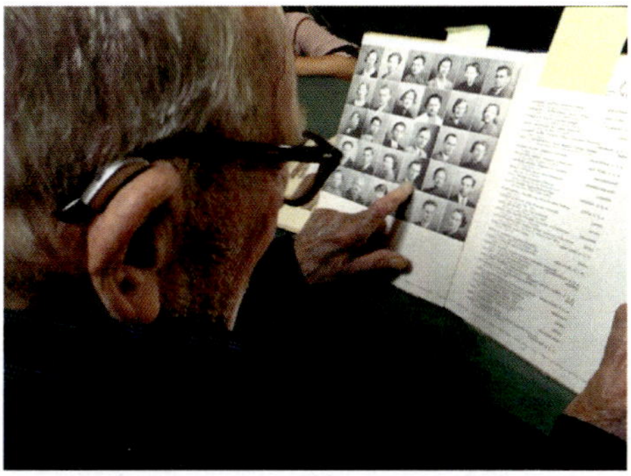

Walter pointing to his picture in the 1936 International House Yearbook. Can you find him in the class photo above?

Walter with Julie Packard from the Monterey Bay Aquarium, catching up on old times following his address to the U.N. committee.

Mainly Mozart "Music and the Mind" Showcase at Seiche.

With Mary at his side, Walter is embraced by His Holiness the 14th Dalai Lama.
Photo: Erik Jepsen/UC San Diego Publications

With Ann Curry following the meeting with His Holiness the Dalai Lama.

Lunch with the von Fuersters, current owners of Eggelgut, Walter's childhood home in Altaussee, Austria.

A visit to the Altaussee Salt Mine. Walter's step-father was Director General of the Austrian Salt Mines.

"Ocean Acoustics, National Security and Walter Munk"
Munk Centennial Symposium at Scripps, 29-30 August 2017.
Photo: Jeff Dillon/Scripps Communications

Walter at Seiche. Photo: Sandy Huffaker

Giuseppe Notarbartolo di Sciara, Walter's student who discovered and named the *Mobula munkiana*, film director and producer Eliana Martinez, and Walter taking a break while shooting "Spirit of Discovery," a documentary about their expedition to Cabo Pulmo.

Poster for the film "Spirit of Discovery." Giuseppe and Walter amidst Munk's Devil Rays in the Sea of Cortez.

★ HONORARY ★

WALTER MUNK WAY

Dedication of the Boardwalk at La Jolla Shores.

Walter's advice on how to live to be 100!

To be continued…

Tribute by Jim Cairns

"There are scientists who are brilliant at applying mathematics in order to advance theories; others are brilliant at conducting complex experiments, still others at devising instrumentation to make the experiments possible, others to analyze data and create empirically based models, others who can brilliantly interpret and communicate scientific results, and so on. But there is a much rarer sort of brilliancy with which very few scientists are blessed. It's the ability to apply a high degree of competency to all of the abovementioned aspects of science, and from that broad perspective understand and communicate the most important and potentially fruitful directions that scientific research should take. You have that rare talent. You have applied it diligently for more than seven decades. You have shared it generously with many others, and you have shown the way forward. The recognition you have received for that is richly merited."

- Jim Cairns, PhD '74, Walter's student and colleague for more than 45 years and Member of the Director's Council at Scripps Institution of Oceanography

Appendix A (2010-2017): Awards and Honors

Crafoord Prize, The Royal Swedish Academy of Sciences (4 May 2010)

Österreichisches Ehrenzeichen für Wissenschaft und Kunst (The Austrian Decoration for Science and Art), The National Council of the Republic of Austria (2010)

50 Years/50 Leaders: One of UC San Diego's 50 alumni leaders who are transforming the world (2011)

Walter Munk Day, Proclamation Honoring Walter Munk by the County of San Diego California (10 May 2011)

Universidad Autonoma de Baja California Facultad de Ciencias Marinas presentation to Walter Munk for participation in Conferencia Magistral "From Exploration to Computer Modeling" (15 May 2013)

Aquarium of the Pacific Ocean Conservation Award (15 June 2013)

Roger Revelle Medal UC San Diego (November 2013)

Explorers Club Medal (15 March 2014) awarded to Walter Munk for his contributions to scientific exploration and discovery

Meeting with Pope Francis (May 2014) at "Sustainable Humanity Sustainable Nature: Our Responsibility" Symposium, Pontifical Academy of Sciences and Academy of Social Sciences

Medal of Honor of the French Navy presented to Walter Munk by General Frachon, the Director of the SHOM (The Naval Hydrographic and Oceanographic Service) at the Assemblee Nationale, Paris (June 2014)

Achievement Rewards for College Scientists (ARCS) Honorary Chair (2015)

Galbraith, Kate, "Walter Munk, The 'Einstein of the Oceans'" New York Times profile (24 Aug 2015)

Global Compassion Summit, University of California, Irvine. Talk on climate change and meeting with the Dalai Lama (6 July 2015)

Aspell, Paula, Interview with Walter Munk National Public Radio, Nova (9 July 2015)

California Legislative Assembly Senate Resolution honoring Walter Munk, and California Senate Certificate of Recognition, presented by Assembly Speaker Toni G. Atkins (19 October 2015)

American Polar Society Medal (3 November 2015)

Revelle College Faculty Fellow (2016)

Oceanology International North America (OINA) honors Walter H. Munk in a special edition of the Catch the Next Wave (CTNW) Conference (6 December 2016)

Walter Munk Honorary Chair Mainly Mozart (2017)

University of San Diego Honors Walter Munk with Doctor of Humane Letters (May 2017)

Walter Munk Way Dedication, The City of San Diego recognizes Walter Munk by naming the La Jolla Shores boardwalk in his honor (18 October 2017)

Walter Munk Day (19 October 2017), Proclamation Honoring Walter Munk by the City of San Diego presented by Councilmember Barbara Bry (18 October 2017)

Appendix B (2010-2017): Bibliography

2010

263a Rosenthal, B. (2010) Dr. Walter Munk: The Father of Surf Reports. Groundswell Publication Annual Publication, 4, 242-255.

263b Munk, W. and C. Pendarvis (2010) Where the Swell Begins. Groundswell Publication Annual Publication, 5, 242-257.

264 Dushaw, B. D., W. W. L. Au, A. Beszczynska-Möller, R. E. Brainard, B. D. Cornuelle, T. F. Duda, M. A. Dzieciuch, A. M. G. Forbes, L. Freitag, J.-C. Gascard, A. N. Gavrilov, J. Gould, B. M. Howe, S. R. Jayne, O. M. Johannessen, J. F. Lynch, D. Martin, D. Menemenlis, P.

N. Mikhalevsky, J. H. Miller, S. E. Moore, W. H. Munk, J. Nystuen, R. I. Odom, J. A. Orcutt, T. Rossby, H. Sagen, S. Sandven, J. Simmen, E. Skarsoulis, B. Southall, K. M. Stafford, R. Stephen, K. J. Vigness-Raposa, S. Vinogradov, K. B. Wong, P. F. Worcester, and C. Wunsch (2010). A global ocean acoustic observing network. J. Hall, D. E. Harrison and D. Stammer (Eds.), ESA Publication WPP-306, Proceedings of OceanObs'09: Sustained Ocean Observations and Information for Society (Vol. 2), Venice, Italy, 21-25 September 2009, Hall, J., Harrison, D.E. & Stammer, D., Eds., ESA Publication WPP-306, doi:10.5270/OceanObs09.cwp.25

264a Munk, W. (2010) Walter Munk Day, *San Diego City Council.*

264b Munk, W. (2010) A Brief Review of My Career. *Hohenems*, 11(2), 18-19.

265 Farrell, W. and W. Munk, (2010). Booms and Busts in the Deep. Journal of Physical Oceanography, 40, 2159-2169.

265a Munk, W. (2010) Roger Randall Revelle, University of California at San Diego Founder's Day.

2011

266 Munk, W. (2011) The Sound of Climate Change: Crafoord Prize Scientific Lecture, *Tellus* 63, 1-8.

267 Worcester, P.E., W.H. Munk, B.D. Cornuelle, M.A. Dzieciuch, and K.E. Wage (2011), Exploring an under-ice ocean cavity with sound, paper presented at XXV IUGG General Assembly, Melbourne, Australia, 28 June – 7 July 2011.

267a Munk, W. (2011) Nansen Center 1986-2011: A View from La Jolla. *Presentations from the Nansen Center's 25 Years Colloquium*, http://www.nersc.no/news/presentations-nansen-center%C2%B4s-25-years-colloquium-available.

2013

268 Munk, W., Miller, G., Snodgrass, F. And Barber, N. (2013) Directional recording of ocean waves from distant storms. *Phil. Trans. R. Soc. Lond. A,* 371, DOI: 10.1098/rsta.2013.0039.

269 Farrell, W. and W. Munk (2013) Surface gravity waves and their

acoustic signatures, 1-30 Hz, on the mid-Pacific [deep?] sea floor, *Journal of the Acoustics Society of America* 134, 3134-3143.

270 Munk, W. (2013), The Perfect Storm, *Proceedings of the American Philosophical Society* 157, no. 4, 1-24.

2014

272a Munk, W. (2014) Munk's Life: Interview by David Dolan, *Explorers Journal,* 92, no. 1, 29-31.

Munk, W. (2014) An Oceanographic Perspective, in Sustainable Humanity, Sustainable Nature: Our Responsibility, a joint Symposium between the Pontifical Academy of Sciences (PAS) and the Pontifical Academy of Social Sciences (PASS), Vatican city, 2-6 May 2014.

2015

273 Mikhalevsky, Peter N., et. al (2015) Multipurpose acoustic networks in the integrated arctic ocean observing system, *Arctic* 68, Suppl. 1, 1-17.

274 Munk, Walter H. (2015) Historical Perspectives: Acoustic Monitoring of Global Ocean Warming: The Very First Encounters with Concern About Marine Mammals and Sound: A Brief Personal Historical Narrative, *Aquatic Mammals* 4, 524-528. http://dx.doi.org/10.1578/AM.41.4.2015.524.

2016

275 Farrell, W. et. al (2016), Wind sea behind a cold front and deep ocean acoustics, *Journal of Physical Oceanography* 46, 1705-1716.

2017

276 Ramanathan, V. et. al (2017) Well Under 2 Degrees Celsius: Fast Action Policies to Protect People and the Planet. Report of the Committee to Prevent Extreme Climate Change.

Hans von Storch · Klaus Hasselmann

Seventy Years of Exploration in Oceanography

A Prolonged Weekend Discussion with Walter Munk

Springer

Hans von Storch
Institute of Coastal Research
GKSS Research Center
Geesthacht
Germany
Center of Excellence CLISAP
University of Hamburg
Germany
hvonstorch@web.de

Klaus Hasselmann
Max-Planck-Institut für Meteorologie
Bundesstraße 55
20146 Hamburg
Germany
klaus.hasselmann@zmaw.de

pb ISBN 9783642426193
ISBN 978-3-642-12086-2 e-ISBN 978-3-642-12087-9
DOI 10.1007/978-3-642-12087-9
Springer Heidelberg Dordrecht London New York

Library of Congress Control Number: 2010922341

Typesetting and production: le-tex publishing services GmbH, Leipzig, Germany
Cover design: WMXDesign GmbH, Heidelberg, Germany

Printed on acid-free paper

Springer is part of Springer Science+Business Media (www.springer.com)

Foreword – Carl Wunsch

Many scientists are impatient with, and uninterested in, the history of their subject – and young scientists are universally urged to focus on what is not understood – to "exercise some imagination" – to look forward and not back. Some understanding of the past is, however, essential for progress. In many fields "what everyone knows" is often merely an accident of the then available technology, or the sometimes arbitrary, specific interests of the small number of strong leaders who set the research directions for decades at a time. In this fascinating interview, Hans von Storch and Klaus Hasselmann elicit from Walter Munk the major elements of a career in physical oceanography, geophysics, and war and peace. The story of Munk's life could readily make a strong character in a novel, although it might be regarded as implausible. Born into a wealthy Jewish banking family in imperial Vienna, he developed (at least as he tells it) into a tennis playing, skiing, ne'er-do-well, exiled to school in rural New York State, following which he did an unsuccessful stint in the family banking business, and migrated to the promised land of California. Employing his undoubted salesmanship skills, he gained a toehold in academia as a Caltech undergraduate.

From there, Walter's life is the quintessential mid-20th century American scientific success story, but with a continuing dollop of drama: enlistment in the U.S. Army, a stint in the Mountain Ski Troops, and a summons back to civilian life just before Pearl Harbor. (His army unit was almost wiped out in the fighting in New Guinea.) His recall led to the hugely important development of the Sverdrup–Munk wave forecasting system and its success in the North African and Normandy landings – but shadowed by unknown, and unexplained for decades, secret accusations of treachery – with resonances in the hysteria of today. Involvement in the atomic and then hydrogen bomb tests followed, with his life-long involvement in JASON and national security issues. After the War, he developed a sure instinct for the most important scientific problems in physical oceanography and geophysics of his era. Both society and the way science is carried out have changed beyond recognition from the inter-war days of Munk's earliest career, to a world of jet-aircraft, near-instantaneous electronic communication, and almost unbelievable (from the vantage point of 1938) transformations in the observational capabilities and understanding of the ocean we now have. Munk's adaptability, his skill in moving with the times over the many decades, and his ability to strike right to the heart of problem after problem, comes strongly through here.

Readers of this oral history will come to appreciate Walter's ability to explain in simple language some of the most esoteric of fluid ocean problems, his sure grasp of observational and theoretical issues, and a sometimes uncanny ability to delineate the essence – that had eluded his predecessors – of a central problem. He has the knack of defining a field in a way that requires decades of subsequent work by others to fully flesh-out, while he himself moves on. One of his explicitly stated themes is that it is more important to ask the right questions than it is to give the right answers. He has managed over and over to do both.

Until the sad loss of Judy in 2006, the expression "Walter and Judy" needed no further explanation for most oceanographers, and for scientists of all flavors, who had encountered this remarkable couple either at home in what seemed like a near-perpetual sequence of lunches, dinners, parties, home theatre productions (Fig. F.1), or in some exotic, but interesting, and one would have guessed, inaccessible-to-a-wheel-chair, city or monument (Machu Picchu, Venice). In their house in La Jolla hangs a double portrait of Walter and Judy back-to-back, facing outward to whatever the world would bring, watching out for each other, for something fun to do together, and ever alert to the well-being of the other (Fig. F.2). There are several stories

Fig. F.1 A dance performance by Jean Isaacs Company at "*The Folly*." Judith Munk designed and built the little amphitheater in the early 1990's in the front yard of *Seiche* overlooking the Pacific Ocean

Fig. F.2 Painting of Judith and Walter by John Asaro (circa 1970)

here, one man's exceptionally interesting life; the transformation of the sciences generally, and of physical oceanography particularly, over the past 70 years; the trajectory of military and national security concerns from World War II right up to today; the tension between hysteria and true security risk.

Hans von Storch, a scientist with a clear understanding of the importance of history to scientific development, and Klaus Hasselmann an eminent collaborator of Walter's, have brought together in this interview the raw material of history. How a young, wealthy playboy immigrant developed into a self-confident dominant scientist of the 20th century, the adviser to generals and admirals, a scientific "wise man" of his generation, while still publishing sharply original science well into his 90s, is an intriguing story eloquently told.

Foreword – Paul Gaffney

Walter Munk. Finally, a book that celebrates this gentle man; a polite, kind and encouraging man. Finally, a book about this man who would look beyond the instant problem and always see the greater issue. Finally, a story that captures Walter Munk, the leader.

Strange. Strange, in commending this long overdue book to you, that I start by describing Dr. Munk's character as a human, rather than simply teasing you with a list of his remarkable scientific accomplishments.

This biography describes a whole man: immigrant, student, father, husband, scientist, oceanographer, teacher, patriot, unselfish mentor, and leader. Young men and women who seek models for the full life should read the chapters ahead.

I was scared to death when first invited to the Munk's home on a bluff splashed by the Pacific's ceaselessness, just up the coast from the Scripps Institution of Oceanography. I had heard about the magnificent home. I had heard about the engaging company and the doubly engaging conversation there. True and true. What I did not expect was the incredible and obvious respect Dr. Munk and his beloved Judy offered each guest in their welcome and in the inevitable discussions about ocean warming, San Diego zoning, respect for other views, partisan politics or any other contemporary topic.

Yes, I stood in a U.S. Navy uniform on the stage in Kyoto with Walter, as his "second" and as bow tie aide, when he received the Kyoto Prize in 1999 (Fig. F.3). What an honor. While the ceremony was inspiring, it was those engaging dinners with the Munks in La Jolla and the many meetings in which I sat with Walter that were the most stimulating.

To science and the Navy. Count the time from early WWII to the first decade of the 21st century; you will see Walter Munk's marks everywhere. We are introduced to his skill in helping to develop wave forecasting in WWII leading to techniques used

Fig. F.3 Rear Admiral Paul Gaffney, Chief of Naval Research, tying Walter's bow tie prior to the Kyoto Prize Ceremony (1999)

on D-Day, then to ocean acoustics and other submarine detection techniques in the Cold War, and recently his attention is drawn to Defense and Intelligence contributions to real understanding of the climate issue. In addressing every challenge, Dr. Munk could see the full breadth of an issue and then zero-in on its essence. Pick the problem, he would easily define the budget burden, the talent requirement and the political overhead, and then he would take out his pencil and start "first principles" calculations. He would offer a few well chosen words, always designed to encourage; never to demean. Yet, when it came to the final analysis of the problem, he was uncompromising with his calculations. He was insistent that the best knowledge, principles of physics, and real data would inform decision making. And, he could make his case by a quick calculation. That is why everyone always wants Walter Munk to serve on every advisory panel.

Finally, as I encourage you to read about Dr. Munk here, I suggest you note his open-minded view of relationships. Here was a man intensely interested in national security issues yet thriving in a liberal environment. I think he saw the two missions: enhancing security in a free nation and gaining knowledge about the processes of our planet as reinforcing, not competing. Here is a true liberal thinker; an open minded man who takes time to think about the big challenges.

Paul Gaffney
Vice Admiral, U.S. Navy (Ret.)
President, Monmouth University
West Long Branch, New Jersey

Preface

It all began with Markus Jochum approaching one of us (HvS) – "when you guys are doing interviews with senior scientists from oceanography and related sciences, why are you not doing Walter Munk?" Indeed, why not? Walter Munk, an icon in oceanography, had just given a wonderful talk in a symposium in honor of his 90th birthday, sweeping a grand circle from his earliest work with Chip Cox on airborne measurements of ocean surface roughness to the latest satellite data – not simply a review, but the struggle of an active scientist opening up new perspectives – as inspiring and stimulating as when one of us (KH) first met him at the Ocean Waves Conference in Easton in 1961 (Fig. I.1). Walter immediately agreed to share with us his recollections on the nearly seventy years of his path-breaking contributions in a sheer amazing range of topics, from ocean waves, internal waves, ocean currents, tides, tsunamis, sea level, microseisms and the rotation of the earth to ocean acoustic tomography.

With "you guys" Markus was referring to HvS and the various partners HvS had invited to join him in conducting a series of interviews of retired colleagues.[1] While HvS was motivated generally in understanding the complex processes by which scientific knowledge was passed from one generation to the next, his colleagues, chosen on the basis of their close professional and personal relationship to the interviewed person, were interested in discovering more about the personalities and motivations that had inspired their mentors.

The conversations with Walter, carried out by interviewers one and two generations his junior, promised to be exceptional in both respects. KH enjoyed four extremely stimulating and formative years from 1961 to 1964 as a young Assistant Professor in Walter Munk's newly established Institute of Geophysics and Planetary Physics

[1] Scientists interviewed by HvS and colleagues: Klaus Wyrtki, Harry van Loon, Reimar Lüst, Klaus Hasselmann and Hans Hinzpeter – see http://coast.gkss.de/staff/storch/interview.htm

Fig. I.1 Taking a break from "The Discussion" (2008). (Susanna and Klaus Hasselmann with Walter at Torrey Pines State Park in San Diego, California)

at the Scripps Institution of Oceanography in La Jolla, California. When he later became the director of the newly created Max Planck Institute of Meteorology (MPIM) in Hamburg, in 1975, Walter, in turn, was a frequent visitor to Hamburg as a member of the Institute's Scientific Advisory Committee. HvS, before becoming director of the Institute for Coastal Oceanography at Geesthacht, near Hamburg, in 2001, was senior scientist at MPIM and had therefore also been exposed to Walter's influence. Both interviewers have been inspired not only by Walter's scientific contributions, but even more by his enthusiastic love of science, and his ability to instill his infectious enthusiasm into everyone he worked with. We have both tried to learn from Walter for the work in our own institutes.

So objective reasons for the interview were there a plenty. And the prospect of sitting on Walter's patio, looking out on the swell running up La Jolla's beaches, talking about old times, the Groundswell Society and new horizons, was an anticipated added pleasure in which we were not disappointed.

But the discussions with Walter also turned out to be exceptional in another respect. What started off as one of Hans' interview projects, developed in the course of editing, as Walter came up with more and more stories and anecdotes, and more and more of Walter's extraordinary scientific career and oceanographic and geophysical explorations surfaced, into a fascinating de facto autobiography. Despite the many things that have already been written about Walter (for example, when he received the Kyoto Prize award in 1999), we thus believe this extensive account in Walter's own words is unique. For the non-scientist, some technical aspects of Walter's many accomplishments may be difficult to fully appreciate, but we believe this personal account of the interweaving of the thrill of scientific discovery with the very human encounters involved in the complex process of discovery will provide equally stimulating reading for the specialist and non-specialist.

Scientific discourse develops like the trunk of a tree.[2] Each year, a new tree ring is formed based on the most recent findings incorporating previous results – from the most recent tree ring, so to speak – and newly established facts and interpretations. Knowledge obtained from earlier research is either encoded or obliterated in present knowledge – continuously transferred from tree ring to tree ring – or forgotten. If something has not been incorporated from cohort to cohort of scientists, it is considered to be irrelevant and of little interest.

Walter's results have long become an integral part of the community knowledge, of accepted concepts, ubiquitous applied terminology – but who apart from family and friends would really care about the person behind it, his perceptions, and his life-long random walk through a maze of inspiring questions about curious phenomena? Is there more to learn than what has ended up in textbooks, such as the dynamics of the western boundary currents? Is it not so that scientific results are getting de-personalized, detached from the inventor, the discoverer, and the assessor? Actually, other scholars later are usually better in explaining, contextualizing and reducing these new results to their essentials, so why refer to the originals, apart from politeness? Yes, this is true, or may at least often be considered a reasonable approximation of reality – but it represents only one aspect of science. Another, equally important aspect is the process of creating scientific knowledge, as opposed to the body of knowledge itself.

Science is a social process, a cultural process, carried out by individuals in a social and cultural context; in a specific time, with fashions, perceptions, limitations, all this. We find it interesting to reflect upon this process; to see the example set by individuals in generating new knowledge, in what is described in German by the wonderfully precise term "Wissen schaffen."

Walter Munk is such an individual. His life illustrates how opportunities, luck, good and bad errors, failures, visions, and intuition are needed just as much as determination, openness and humor for making progress. Walter is a prototype of a scientist – a model in terms of results, performance, and humanity. A man who is proud of his successes and aware of his status, but whose modesty, inborn courtesy, generosity, and humor disarm and charm everyone he encounters. Through his open personality he has inspired many younger scientists, while his recollections reveal that he in

---

[2] Stehr, N., von Storch, H. (Eds.): *Eduard Brückner – The Sources and Consequences of Climate Change and Climate Variability in Historical Times*. Kluwer Academic Publisher (2000)

turn was equally strongly influenced and motivated by the leading oceanographers and geophysicists of his time.

Walter's life was no doubt conditioned by the big social upheavals in the Old World; first the Nazis, who ensured his staying for good in the U.S., becoming early a U.S. citizen, and a member of the U.S. armed forces; secondly the Communists, the fight against which he supported by responding to the needs of the Navy and as a member of JASON, by advising all U.S. administrations since Kennedy on science and issues of national security. These issues appear throughout the interview and are discussed in detail in the section on the "Navy."

Walter's account is also interesting with respect to the role of science in society. The philosophy was set in 1945 by the seminal report *Science – The Endless Frontier* by Vannevar Bush, according to which the researcher was obliged to "produce and share knowledge freely to benefit – in mostly unspecified and long-term ways – the public good."[3] More specifically, "The centers of basic research... are the wellsprings of knowledge and understanding. As long as they are vigorous and healthy and their scientists are free to pursue the truth wherever it may lead, there will be a flow of new scientific knowledge to those who can apply it to practical problems in government, in industry, or elsewhere."[4] Small basic-science groups, with adequate funding, occasionally brought together in larger ad-hoc coalitions, created the "new knowledge," not big programs, with detailed a-priori project planning, grandiose but often empty promises, milestones and deliverables, as we know them from monstrous European-Framework projects. Vannevar Bush's concept stipulated a linear model of the relationship between science and society: basic → applied → development → application → societal benefit, as a more or less automatic assembly line.[5]

The Vannevar-Bush-doctrine was valid throughout most of Walter's career, but was suspended in the late 1990s, as the call for change in an assessment of the U.S. House illustrates, "*The United States has been operating under a model developed by Vannevar Bush in his 1945 report.... It continues to operate under that model with little change. This approach served us very well during the Cold War, because Bush's science policy was predicated upon serving the military needs of our nation, ensuring national pride in our scientific and technological accomplishments, and developing a strong scientific, technological, and manufacturing enterprise that would serve us well not only in peace but also would be essential for this country in both the Cold War and potential hot wars. With the collapse of the Soviet Union, and*

[3] Office of Technology Assessment: *Federally funded research: decisions for a decade*. Washington, DC: OTA-SET-490, U.S. GPO (1991).

[4] Bush, V.: *Science – the endless frontier*. Government Printing Office, Washington, DC (1945)

[5] Pielke, R. Jr.: *The honest broker*. Cambridge University Press (2008)
Modern science studies have informed us that this assembly line is a piece of myth, and does not adequately describes the interaction of science and society.

the de facto end of the Cold War, the Vannevar Bush approach is no longer valid."[6] At that time, Walter was 80, so it did not need to concern him.

Walter Munk belonged to a fortunate generation of scientists who could mostly determine themselves how to pursue their science. As long as it was considered "good science," everything was fine, because – in the logic of Vannevar Bush – this would eventually lead to useful results, useful for the nation, for the military, for industry. Certainly, not all projects would succeed in this respect, but a sufficiently large percentage of all projects. Thus he is telling us about an approach to science that deviates markedly from contemporary top-down approaches, where it is often the number of publications in *Nature* and *Science* that decide on tenure and careers. Maybe the old model of science is irrevocably gone; maybe there will be some kind of renaissance sometime later.

There are also other things to learn from Walter. He is not talking about university committees, about the need to be present when the purportedly important people meet, when allegedly significant decisions are taken. Instead, he just goes on doing science, thinking about waves, tides, the rotating earth, sound propagation in the ocean – always on the move, from the short term to the long term, always inventing new terminology, always grasping the hidden core of a problem. We have seen many

Fig. I.2 Judith and Walter celebrating their 50th wedding anniversary on the patio at *Seiche*, their self-designed and self-built home north of Scripps Institution of Oceanography (2003). In Austria, Walter's homeland, country houses were often named. Seiche is an oceanographic term for oscillations of a lake or landlocked sea, however there is no scientific significance here; Judith's mother chose the name because she liked the sound of the French word. Judith and Walter have welcomed guests from around the world to *Seiche* for over 50 years

[6] U.S. House of Representatives Committee on Science: *Unlocking our future: towards a new national science policy*. Government Printing Office, Washington, DC (1998)

colleagues who spent the last ten years or so of their career building empires, in asserting importance, who disappeared soon after their inevitable final retirement into oblivion because of lack of substance. For Walter, the issue was always knowledge, and the generation thereof, and not power.

But, there are also characteristics that one cannot learn from Walter, which are simply there: talent. Walter has an uncanny instinct for problems that are ready to be solved. And he invariably drives to the core of a problem, invents new terminology to capture the essence of the solution, and orders the new insights into a remarkably clear mental record of the overall picture. In contrast to, say, Hilbert, who is alleged to have forgotten his seminal publications of only a few years ago, Walter has instant recall and can give a print-ready explanation of all the phenomena on which he has worked – to our continual amazement during the interview.

And, of course, another fact one cannot learn from Walter but only admire: his good fortune in being married to Judith (Fig. I.2), who supported him throughout his career, both with her courage, optimism, and empathy with people, and in her active contributions as creative architect in designing the Institute of Geophysics and Planetary Physics (IGPP) buildings.

Fig. I.3 Walter's philosophy in teaching and life has always been "Keep IT Simple!" and "MAKE IT FUN"!

Finally, and above all – we are not sure that this is something that one can learn or simply has to be infected with – Walter has an immense fun in whatever he does. This is surely one of the reasons that he is a happy and joyous man at his youthful 92 years (Fig. I.3).

Acknowledgements

Andrea Santos has patiently assembled the diverse information going into this account. We are deeply indebted to her. We are also deeply indebted to Deborah Day, Archivist Emeritus of the Scripps Institution of Oceanography for correcting a large number of inaccuracies in this free-wheeling discussion. Those inaccuracies that remain are our responsibility, not Deborah's. Thanks to Ilona Liesner and Beate Gardeike from GKSS for their editorial support.

The Office of Naval Research, United States Navy, has generously supported the publication of this volume. Walter Munk has held the Secretary of the Navy Chair in Oceanography at the Scripps Institution of Oceanography since 1985.

Contents

Acronyms

AFOSR	Air Force Office of Scientific Research
AMSOC	American Miscellaneous Society
ASW	Anti-Submarine Warfare
ATOC	Acoustical Thermometry of Ocean Climate
BOMM	Bullard, Oglebay, Munk, and Muller (computer program)
CA	Coarse Acquisition
CalCOFI	California Cooperative Oceanic Fisheries
CAT	Computed Axial Tomography
Cdr.	Commander
CSIRO	Commonwealth Scientific and Research Organization
CUSS	Continental, Union, Superior, and Shell Oil Companies
CW	Continuous Wave
DAMTP	Department of Applied Mathematics and Theoretical Physics
DSDP	Deep Sea Drilling Project
FBI	Federal Bureau of Investigation
FTE	Full Time Equivalents
GM	Garrett-Munk
GPS	Global Positioning System
GSI	Geophysical Services Inc.
GW	Gigawatt
HMAS	Her Majesty's Australian Ship
HO	Hydrographic Office
ICES	International Council for the Exploration of the Sea
IDA	International Deployment of Accelerometers
IGP	Institute of Geophysics
IGPP	Institute of Geophysics and Planetary Physics
IPCC	International Panel of Climate Change
IWEX	Internal Wave Experiment
JASA	Journal of the Acoustical Society of America
JONSWAP	Joint North Sea Wave Project
LCVP	Landing Craft Vehicle and Personnel

lod	length of day
MIT	Massachusetts Institute of Technology
MOC	Meridional Overturning Circulation
MODE	Mid-Ocean Dynamic Experiment
MPIM	Max Planck Institute of Meteorology
NAS	National Academy of Sciences
NOAA	National Oceanic and Atmospheric Administration
NSF	National Science Foundation
OAT	Ocean Acoustic Tomography
ODI	Ocean Desing Inc.
ONR	Office of Naval Research
R/V	Research Vessel
RLC	An electrical circuit consisting of a resistor (R), an inductor (L), and a capacitor (C), connected in series or in parallel
SCOR	Scientific Committee on Oceanic Research
SOSUS	Sound Surveillance System
TLP	Tension Leg Platform
UCDWR	University of California Division of War Research
UCLA	University of California at Los Angeles
UCSD	University of California at San Diego
USS	United States Ship
WHOI	Woods Hole Oceanographic Institution

Chapter 1

Waves and Wave Spectra

1.1 Oceanographers Learn About Power Spectra

Hasselmann: Many of the topics you spearheaded in oceanography and geophysics were driven by your knowledge of time-series analysis, of spectral analysis and so forth. You were one of the first people to develop a general spectral analysis program, the BOMM (Bullard, Oglebay, Munk, and Miller) system [91]. *And you applied that program to explore the ocean wave spectrum. Could you say a few words how you came to become interested in applying these spectral analysis techniques to oceanographic phenomena?*

Munk: I first became aware of the problem in connection with wave prediction. During the war a remarkable UK group under George Deacon (later Sir George) had developed an analog method of getting power spectra for waves. Norman Barber from New Zealand did the pioneering work. He recorded on a film half black half white, with a wavy boundary that represented the wave record. The film was viewed through a vertical slit by a photocell whose output was the fraction of white film (the wave elevation). The film was mounted on a spinning wheel that was allowed to slow down by friction. The output of the photocell through a resonant RLC-circuit was recorded on paper. So as the wheel spun down, lower frequencies got into resonance. These were the first power spectra of ocean processes.

Hasselmann: I find this hard to believe. The acousticians knew about power spectra long before then.

Munk: Indeed they did. So did the optical people. They were monitoring in frequency space (pitch and color) in the first place. But I believe there was not a single oceanographer that knew how to handle time series of random phase processes. There were a few clumsy attempts to apply tidal analysis to represent wave records by a few fixed-phase frequencies. We oceanographers depended on Barber's wheel to lift the low wave frequencies of order 0.1 Hz by four order of magnitudes to the 1000 Hz resonant frequencies of RLC-circuits.

H. von Storch, K. Hasselmann, *Seventy Years of Exploration in Oceanography*
DOI 10.1007/978-3-642-12087-9,

But we were not the only ones who could not handle phase-incoherent processes. Meteorologists were in the same boat. So were seismologists, as we learned rather painfully during the test ban negotiations. So were geologists. But for the geologists random phase wave analysis was not such a useful tool (unless you think of ripples on the sea bottom). It is not fruitful to think of Yosemite's Half Dome as a random-phase process.

Hasselmann: And how did this eventually get resolved?

Munk: With the advent of digital computers. The earliest attempt was to digitize records and form autocorrelations and their cosine transform [67]. I owed all this to John Tukey, who took me under his wing; John was a Princeton Professor, an active consultant to the Bell Laboratories and a leader in this field. Early on he appreciated that we needed to find a good digital method of getting spectra of random-phase processes. John Tukey was fascinated with new applications and he came out to La Jolla and stayed at our house. At the time (I think 1955) the preferred method was to obtain power spectra by computing cosine transforms of the auto-covariance. (A few years later Tukey was to pioneer the fast Fourier transforms now universally used.) It was during the early days of IBM punch cards, and we wrote a program in machine language and put some wave records on punch cards. Convair, a local airplane company, offered the use of their IBM 650 between 03:00 and 06:00 at no cost. I would get my cards ready at night and drive to downtown San Diego at three in the morning and return at 6 AM with a power spectrum. By 1965 we had

Fig. 1.1 Kendall, Edie, and Judith on the patio at *Seiche* (1961)

advanced to eight-hole tape, to the delight of my daughters who spread kilometers of black tape throughout our canyon (Fig. 1.1).

I must tell you a related story. Some years later I went to University of California at Los Angeles (UCLA) to compute wave spectra and there was the great man Norbert Wiener doing analyses of brain waves. Someone suggested that Wiener would be interested in what I was doing and I was consequently introduced. Wiener said, "What are you doing, Munk?" and I replied, "I'm computing spectra of ocean waves." "What method are you using?" I had learned that he was a somewhat vain man, so I said, "I am using the Wiener method." He said, oh, tell me in detail how you proceed. I said, "Well, we digitize a wave record $x(t)$, so we get $x_1, x_2, x_3 \ldots$" Wiener interrupted, "You are not taking enough data." He was of course right.

1.2 Wave Prediction

Hasselmann: So your entry into oceanography was in ocean waves. Could you go back further and tell us how you started in that area?

Munk: I think you are referring to the time when I was working on wave prediction for the Navy. I already had a master's degree in oceanography; I had spent two summers at Scripps Institution of Oceanography. Then I had served in the U.S. Army for 18 months. My real introduction came when I was working in Washington at the Pentagon on various oceanographic problems. I learned that the allied forces, England and America etc. were planning for a landing in Northwest Africa. This was 1943, about two years after Pearl Harbor. For two years the Allies had responded to Axis initiatives; this was to be the first time that the initiative was on our side.

I asked for permission to witness the preparations. I was permitted to go to the Carolinas and observe some practice landings. The principal equipment was a landing craft called LCVP (for Landing Craft Vehicle and Personnel). LCVP's are still in existence. These are small landing craft that come into the beach and lower the bow, and people jump out and climb up on the beach. The landing craft was of such nature that if the waves were too high, the craft would tend to broach, turn parallel to shore, and the waves would break over the side and fill the landing craft with water. People would get hurt and the practice landings would be postponed until some later days when the waves were lower. The cut-off height was something like five feet. I returned to Washington to learn something about wave statistics in Northwest Africa. It was clear that during the winter, the proposed landing time, waves were generally higher than five feet. It seemed like a catastrophe about to happen.

Hasselmann: And how did you respond?

Munk: The obvious response was to learn how to predict waves so that one could pick a couple of calm days. As far as I could tell there had been no previous attempt at wave prediction. The indicated procedure was to divide the subject into three parts: Sea, Swell, Surf. Sea consisted of computing wave dimensions, height and period, as a function of wind speed and storm "fetch," using weather maps. Swell was to estimate the attenuation as the waves traveled from the storm area to the landing beaches. Surf was to estimate the amplification and transformation of the waves as they came into shallow water.

I spent some weeks learning about the problems and then reported to my superior, Major H.R. Seiwell (an Army Air Corps officer who had studied at Woods Hole), that we should make a major effort at wave prediction. Otherwise the Allies could face the potential of a major disaster. I was twenty-five years old with no track record, and I was dismissed with the comment that "they" must have taken all this into account. (In the intervening sixty or so years I have yet to find out who "they" were.) I telephoned Scripps Director Harald Sverdrup who came on the next plane. We spent a month working on the problem. He concluded that it could be done. Sverdrup was a world-renowned oceanographer and polar explorer, and his advice was followed. We worked out a crude method, very crude, based on empirical published data, mostly using dimensional consideration. Eventually two relatively calm days were picked for the landing. The landing was successful.

There was a large element of luck. Prior to the landing we were desperate to test the prediction formalism. It turned out that Pan-American Airways had landed seaplanes in the Azores and had kept a record of wave heights. Using historical weather maps we "hindcast" the Azores wave conditions. Things looked pretty good but once in a while high waves were recorded which were totally missed by our hindcasts. I noticed that the "spikes" were curiously spaced at even intervals. On further examination the spikes invariably occurred on Saturday nights. We allowed ourselves the interpretation that they were more nearly related to Portuguese vin rosé than to meteorological conditions.

Hasselmann: What happened in relation to amphibious landings for the rest of the war?

Munk: We started a school for wave prediction here in La Jolla. Our students were Navy and Air Force (originally Army Air Corps) weather officers. We must have graduated about a hundred officers. Some of them became leaders in future oceanography. We learned as we taught, the prediction method was pretty much transformed from one class to the next. The officers did the predictions for all amphibious landings in the Pacific Theatre of War, with Sverdrup and me playing diminishing roles. Eventually in collaboration with the British Met Service a prediction was made for the Normandy landing, where it played a crucial and dramatic role. The prediction for the first proposed day of landing was that it would be impossible to have a successful landing. As I understand, the wave prediction persuaded Eisenhower to delay for 24 hours. For the next day the prediction was "very difficult but not impossible." Eisenhower decided not to delay the second day, because the secrecy would be lost in waiting two weeks for the next tidal cycle.

After the war the Hydrographic Office (HO) published the method as HO 601 and HO 602 [2, 3]. We improperly had called HO 601 a *theory* for wave forecasting. It was not a theory. It was a collection of recipes aptly assembled. Klaus, you were kind enough to give a talk entitled "HO 601" at my 65th birthday.

Hasselmann: I remember the following statement in Blair Kinsman's book on Wind Waves, "... there are some thousands of World War II veterans alive today who would have been dead in the surf had Sverdrup and Munk not done their best with what they had."

Yes, and I well remember the 65th birthday day surprise symposium that your wife Judy sprang on you in 1982. I still see the surprise in your face when you opened the door of the Sumner auditorium to "meet somebody" and were greeted with a full applauding auditorium! In my talk I presented the new wave model we had developed following our 1969 Joint North Sea Wave Project (JONSWAP) on wind-wave growth (that was inspired by your pioneering swell measurement project in the Pacific that we shall come to later). I showed a transparency of predicted wave height versus fetch and said I would now compare this with the HO 601 results. But I warned that this would be hardly fair. After all, the new wave model was calibrated against 13 wave instruments recording wave spectra along a well defined profile under ideal off-shore winds, and was based on a spectral theory of wave growth, including John Miles' theory of wave generation and the important but highly complex nonlinear spectral transfer integral – all of which nobody had even dreamed of when you created HO 601 by applying dimensional analysis to a hodge-podge of visual wave observations, including sources as diverse as the ripple waves in a Hyde Park pond. I believe the burst of applause when I overlaid a transparency of the HO 601 data showing a perfect fit with the 2nd generation wave model and the JONSWAP data was a nice birthday present.

von Storch: May I ask one detail that I always wanted to know. Where does the term "significant wave height" come from?

Munk: I think that we invented that word, and it came about as follows: after returning to Scripps we were monitoring practice landings under various wave conditions by the Marines at Camp Pendleton, California. After each landing we would ask the coxswains of the landing crafts to estimate today's wave height, 7 feet one day, 4 feet the next day. We would make simultaneous wave records, and compute root mean square (rms) elevations. It turned out that the coxswains' wave heights far exceeded twice the rms elevations. It was easier to define a new statistical quantity than to modify the mindset of a Marine coxswain, so we introduced "significant wave height" (rms of the highest one third waves) as being compatible with the Marines' estimates. To our surprise that definition has stuck till today.

von Storch: Since when are ships of opportunity recording waves? I mean they routinely report significant wave height in their logbooks, don't they?

Munk: Yes, of course. I believe their estimate is near to that of the Marine coxswains. And so are the estimates of surfboard riders.

von Storch: What is a coxswain?

Munk: He is the person with his hand on the throttle looking out to sea and, at the moment he considers favorable, guns it into the beach. This involves an intuitive but very sophisticated ten-second wave prediction. Many coxswains have an uncanny ability to pick the right moment.

von Storch: So, the technical term of significant wave height is from your pen.

Munk: As far as I know. I might add that this is not a great scientific accomplishment.

Hasselmann: But this generation no longer thinks about what it means. So, it is good to hear your explanation of how it came about.

Munk: There is a Scripps Wave Report on significant wave height somewhere in our archives. The development of a wave generation theory was to be left to O.M. Phillips, Miles and Hasselmann.

I should mention a brief consulting interlude. My return to Scripps coincided with the earliest attempts at offshore drilling for oil. I received a contract from Humble Oil to compute wave forces on an offshore drilling platform to be built in the Gulf of Mexico [10]. The platform was built and subsequently demolished in a hurricane. An oil journal (I have been unable to identify it) appeared with a cover, to the left showing the Humble platform after the storm (what was left of it), to the right a "Table of forces computed by Professor Munk." That was the end of my career as a wave consultant.[1]

1.3 Where the Swell Begins

von Storch: So you were engaged in this wave business in the 1940's. Did you follow that business closely in the next 50 years? Or did you leave it mostly to others?

Munk: I made a conscious and deliberate attempt to get away from sea and swell. And indeed our major effort was directed towards waves of increasingly longer periods, culminating in a decade of studying tides. But this attempt never quite succeeded. The measured sloping "ridges" in f, t-space were too dramatic to be ig-

[1] But I remained in contact with problems of wave forces on off-shore structures through my brother-in-law, Edward Horton. Ed developed the first tension leg platform (TLP), moored off California in the early 1970's, and later the deepwater offshore production facilities called spars. Ed's engineering degree at Yale University had not prepared him for dealing with random processes like ocean waves, and I passed on to him what I had learned from John Tukey. Working with Ed has been one of the pleasures of my career.

nored. Let me start with a very recent telephone call, "Congratulations, you have been made an honorary member of the Groundswell Society."

von Storch: the Groundswell Society? What is that?

Munk: I had never heard of it. It is a group of several thousand Californians who enjoy surfing together. And to my regret I have never been a competent surfer. And when I asked why I had been so honored, the caller said, "for having discovered the source waters of the California summer swell." They meant "source waters" in the sense that Victoria Falls were once discovered to be the source waters of the Nile river.

The source waters of the California summer swell are in the winter storms of the Southern Hemisphere. This was not known at the time I returned to La Jolla after the war. It is now common knowledge among the thousands of enthusiasts who ride the waves on the California beaches every summer. A Web site tells them, "a 6 foot swell of 16 s period from west of New Zealand (direction 205 degrees true) will arrive Saturday, expect good breakers at Wind-and-Sea beach in La Jolla," etc.

Would you like me to talk about how the source waters were discovered?

von Storch: Please, do so.

Munk: It is the result of three expeditions. In 1956 Frank Snodgrass and I were taking measurements of pressure on the bottom in 500 feet of water off Guadalupe Island, a Mexican island to the south of Ensenada. The experiment was unrelated to ocean swell. But then we recorded an astounding sequence of dispersive wave signals from a distant source [57]. The wave train is initially of very small amplitude, millimeters, but easily detected by spectral analysis. The sequence was one of gradually decreasing wave period starting with 19 s waves, followed in four days by 14 s. It is possible to compute the sourced distance from the rate of decrease of wave period, using 19th century wave theory. The result was astounding: 13,500 km. Some later events came from even further, approaching the antipodal distance of 20,000 km. The Pacific Ocean is large, but not that large. We returned to La Jolla puzzled and confused.

von Storch: And what was your next expedition?

Munk: There was one way of reconciling the most distant ranges with geography: namely that the source storms were in the Indian Ocean and that they entered the Pacific along great circles through the narrow 220–225 window between New Zealand and Antarctica or the 230 to 236 window between New Zealand and Australia (the Tasman Sea). Radio Astronomers had just pioneered determining the direction of radio stars by using multiple antennas and measuring the phase *difference* of the incoming radio waves. Here was the opportunity of extending the power spectral analysis to cross-spectra between two records. Frank Snodgrass established a triangular 100 m array in 100 m of water on the western (exposed) side of San Clemente

Island (100 km west of San Diego). Norman Barber (of the "Barber Wheel") came over from the UK to help with the analysis [85]. And there we were; waves from the distant storms came in through the narrow window between New Zealand and Antarctica! For one particularly intense event we felt confident to "invert" the wave data to sketch a weather map of the source region. It showed a storm centered on an island called Heard Island in the Indian Ocean at 52 degrees south.

von Storch: So that's where the swell begins. You mentioned a third expedition.

Munk: Having established some of the great circle wave paths, we thought that it would be interesting to occupy a set of stations along such a path [96]. We chose six stations: New Zealand, Samoa, Palmyra (an uninhabited equatorial island), Hawaii, the Scripps' ship FLIP, and Yakutat, Alaska. I took Samoa (Fig. 1.2), Klaus took Hawaii; Gaylord Miller (our only graduate student) took Alaska. We were able to follow wave disturbances for 10,000 km all the way from source to finish.

von Storch: What else did you learn?

Munk: By then Klaus had done the pioneering work on wave-wave scattering, and we spaced the stations to measure the scattering of the southern swell by the trade

Fig. 1.2 Measuring ocean swell from a Fale in Tutuila, American Samoa (1963). Walter had persuaded Judith to take the 0400 daily watch of swell recording. (Edie, Kendall, Judith, Walter, and Silau playing the guitar)

wind sea. This turned out to be negligible; all measurable scattering occurred before the waves left their generation area. So in retrospect the stations were not well spaced. But we could place an upper limit on attenuation. Curiously only now, more than fifty years later, our attenuation estimate of 0.2 dB per 1000 km was useful in some new work on nonlinear wave interactions.

Hasselmann: Indeed, we have come a long way in the wave forecasting business since the first large-scale multi-wave-station measurements you initiated in the Pacific swell experiment. The European Centre for Medium Range Weather Forecasts, for example includes in its routine medium-range (10 day) global weather forecasts global predictions of two-dimensional (frequency-directional) wave spectra on a 0.25×0.25 degree grid. These are based on detailed computations of the wave spectral energy balance, as had been applied already in the analysis of the Pacific swell data in the near-storm regions, and then later in greater detail in the follow-up Joint North Sea Wave Atmosphere Program (JONSWAP) wave-growth experiment. The very weak attenuation of swell outside the generating region found in the Pacific swell experiment is an important factor in the present 3rd generation wave prediction models, and is well supported by modern global satellite measurements of wave heights (from radar altimeters) and two-dimensional wave spectra (from synthetic aperture radars).

Munk: We never dreamt that wave prediction would become a thriving business. We did however realize that the Samoa wave measurements would permit a pretty good job of predicting Hawaii waves a few days later, and for a moment might have entertained the idea of going into the wave forecasting business. However at that time the economic standing of the surfing community did not appear a proper basis for our future economic well being.

Perhaps this is a good time for paying tribute to my partner Frank Snodgrass. Frank was a superb ocean experimentalist; I am not a good experimentalist (Fig. 1.3). But we shared interest in exciting problems and worked together from 1956 to 1975. We published together. Frank pioneered dropping instruments to the sea floor and recalling them some months or years later by acoustic commands.

After Frank died, I started a partnership with Peter Worcester, much of this dealing with ocean acoustics. This has by now lasted thirty years. When we have gone to sea together, Pete has assigned me various duties that keep me from interfering with his instruments. It is a simple fact that the acoustic work, and the work on waves and tides would not have happened without these partnerships.

Hasselmann: I find your account of how you became involved in ocean waves fascinating, beginning with the urgent need to forecast waves for the war effort, and progressing into understanding how waves propagate, and the details of their spectral properties and energy balance, etc. I find it fascinating also because it parallels similar, apparently largely independent developments that you mentioned in England. Fritz Ursell and Norman Barber had also deduced, that swell originated in distant

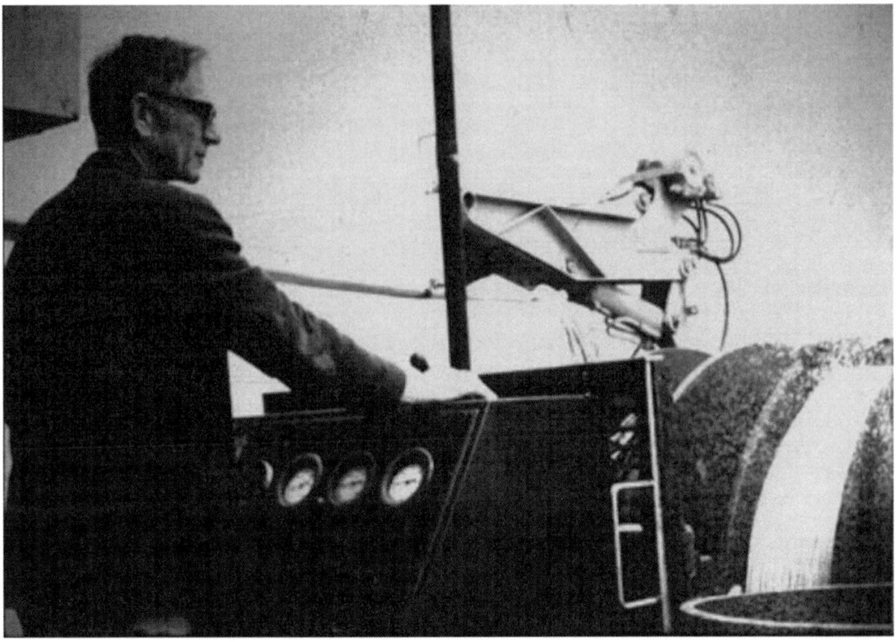

Fig. 1.3 Walter running the winch on one of the rare occasions Frank Snodgrass permitted him. Photo by Don Altman

storms, in this case in the Atlantic. They inferred this by applying wave propagation theory to successive measurements of wave spectra they had made in South Cornwall, I believe in 1944. The peak periods were found to gradually decrease with time, and from the rate of decrease they inferred the distance of the source region. And the motivation of the British wave group, under the leadership of George Deacon, was also to improve wave forecasts during the war. How was the interaction with the British colleagues? You mentioned that Norman Barber visited Scripps in the sixties ...

Munk: At one time or another, we have collaborated with nearly all members of the Deacon Group, with Michael Longuet-Higgins, Barber, Ursell, M.J. Tucker, and with Sir George himself.

Chapter 2
Coming to America

von Storch: You were born in Austria. I am puzzled how that relates to your career in oceanography.

Munk: It does not; Austria is a landlocked country. But don't forget, at one time we owned the North Adriatic. When I started in La Jolla in the late 40's, Austria had a leading school of climatology. Albert Defant was Professor at the University

Fig. 2.1 Since learning to paddle a Plättel in Alt-Aussee as a boy, Walter had been hoping for an opportunity to be his own gondolier on the Venetian canals. It finally came while he and Judith were in Venice on a sabbatical testing new technology to clean marble statues with laser pulses (1972). Walter also spent time working on the design of the Gates to keep aqua alta out of Venice, a 7 billion project that is now nearing completion over 36 years later. At the same time, concern about the increasing lagoon pollution prompted Walter to write a paper (unpublished) "Let the Moon Flush the Lagoon" which advocated timing the opening and closing of the Gates to flush the lagoon by rectifying the lagoon circulation

DOI 10.1007/978-3-642-12087-9, © Springer 2010

Fig. 2.2 Much of Walter's boyhood, prior to coming to America, was spent at the Eggelgut, a 17th century peasant house renovated by his grandfather, Lucien Brunner. It was located on a steep meadow between the forest and a brook in Alt-Aussee, a village about 45 minutes out of Salzburg. Life was centered around the lake and tennis courts. In the winter the family skied the Loser Mountain on a run that ended right at the house. After the war, Walter's mother sold the Egelgut, but kept some land by the lake. Judith and Walter visited often and dreamt of converting the boathouse into a summer residence

of Innsbruck; his two volumes on Physical Oceanography are among the classics, almost in a class with *The Oceans* by Harald Sverdrup, Martin Johnson, and Richard Fleming. Defant spent a year in La Jolla in the mid-fifties as guest of Dr. Louis Lek in La Jolla while Lek translated the second volume into English (published with the support of the Office of Naval Research). I vividly remember being shown a curve supposedly representing a wide scatter of points. When I looked puzzled, Defant replied in his best Tyrolean accent, "Man muss doch einen kosmischen Schwung haben.[1]"

Years later in 1981, when we lived in Venice, a Professor at the University of Trieste took us to his apartment to show us his collection of oceanographic instruments (Fig. 2.1). And there was something very much like the Nansen bottle with reversing bottles and reversing thermometers, built in Vienna prior to the turn of the century. A cable with a string of bottles attached would be lowered to the desired depth, and then raised a few meters. The bottles are tripped by the reversal in *vertical* velocity measured by propellers attached to the bottles (not so good in a heaving

[1] But one needs to have a cosmic swing.

Fig. 2.3 Walter and his grandsons (Lucien, Maxwell, and Walter) in Alt-Aussee facing the Trisselwand (1995)

sea). I believe the Norwegian Nansen bottle was developed after the turn of the century. Here the tripping is accomplished by messengers sliding down the wires.

2.1 Coming to America

von Storch: How in the world did you make your journey from a village in Austria to the beaches of California?

Munk: In a set of steps resembling a random walk. I will have to go back. I grew up in a drafty formal house in Vienna (now the Embassy of South Korea). We spent three months each summer and Christmas holidays in Alt-Aussee (near Salzburg) at the Egelgut, a charming 18th century farmhouse rebuilt by my grandfather (Fig. 2.2). We all considered the Egelgut as home. It is on a steep hillside, bounded by the Egelbach at the lower boundary, and at the upper end the Egelwald, leading up the Loser Mountain. After the war ended I took my family to Alt-Aussee on a number of occasions (Figs. 2.3 and 2.4).

My maternal grandfather Lucian Brunner was a banker, later a member of the Vienna city council (Fig. 2.5). In his later year he turned to Socialism, and changed his bank from Bank Lucian Brunner to Österreichische Volks Bank (Austrian Peoples Bank) but kept all the shares. Lucian came from a Jewish family in St. Gallen,

Fig. 2.4 Family portrait taken in celebration of Judith and Walter's 50th wedding anniversary (20 June 2003). (*Standing*, *left* to *right*: Walter, Lucien, Walter, and Edie; *Seated*, *left* to *right*: Maxwell, Kendall, and Judith)

Switzerland, later settling in Hohenems, Austria (there is still a Brunner Strasse and a Brunner Haus). An Italian branch settled in Trieste where they turned Catholic more than a hundred years ago. My Italian cousins come to Zöbern, Austria every August to be "in residence" at Schloss Ziegersberg (30 rooms with private chapel). My parents (Fig. 2.6) were divorced when I was ten and mother married Rudolf Engelsberg (Fig. 2.7) who was a sub-cabinet member of the Schuschnig government when Hitler invaded Austria. I was brought up in a non-religious household.

Mother's brother Felix was an early glider enthusiast and an avid skier, and first took me up the Loser ("house mountain" of the Egelgut) when I was five years old. Uncle Felix had gone skiing with Hannes Schneider who transformed "modern" skiing from the Telemark school to the Arlberg school. I loved skiing and disliked high school (Fig. 2.8). At fifteen I decided on a career as ski teacher. Mother was appalled. One evening we had an American guest for dinner who told mother that he had a son just like me, and that he had found the ideal school where boys learned the habits of discipline and hard work. So I was exiled to the Silver Bay School for Boys on Lake George, N.Y. A year later we had started the Silver Bay Ski Club and I was president.

Fig. 2.5 Walter's maternal grandparents, Malwine and Lucian Brunner. Lucian was owner of Bank Lucian Brunner. When he became a member of the Vienna City Council, he changed the name to Österreichische Volksbank (Austrian Peoples Bank), but kept all the shares

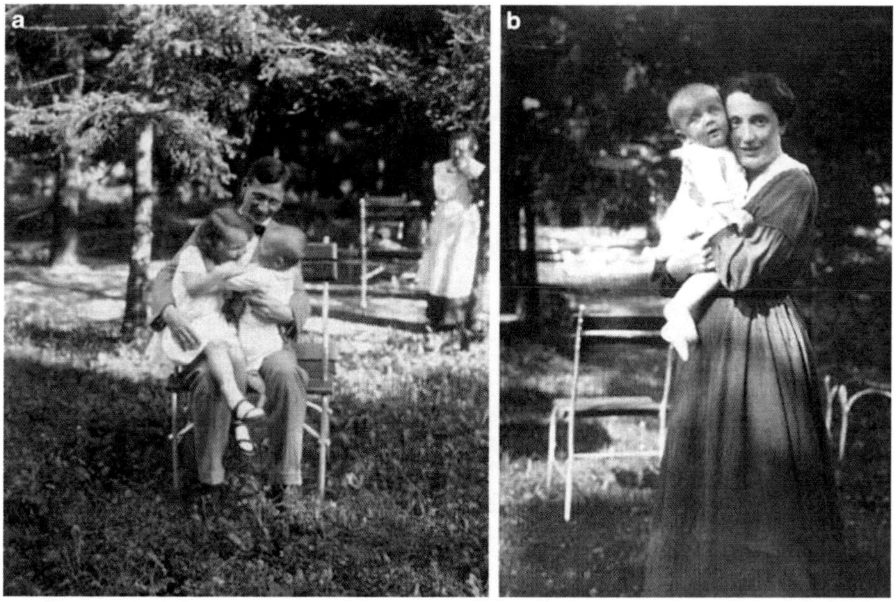

Fig. 2.6 (**a**) Walter with his father, Dr. Hans Munk, and his sister, Gertrude, in their Vienna garden. (**b**) Walter with his mother, Regina Brunner (circa 1918)

Fig. 2.7 Walter's parents were divorced when he was ten years old. His stepfather, Rudolph Engelsberg (*right*), was General Director der Österreichischen Salinen (President of the Austrian Salt Mines)

Fig. 2.8 Mixed Doubles at the Brunner's tennis court in Alt-Aussee (1932). Walter (*right*) took tennis very seriously and once made it to the Austrian Juniors Doubles semi-finals. He and his mother enjoyed competing in mixed-doubles tournaments. Walter's youth was spent skiing and playing tennis, with no signs of any intellectual curiosity

Grandfather Lucian had been in partnership with Cassel & Co in New York; my uncle Julian Triar was senior partner. Upon graduation I became a runner at Cassel's to learn the business from the bottom up. I hated every minute of it. I had the good sense to enroll in night school at Columbia University. Two years later I became twenty and mother was ready to give up on her hopes for Walter's banking career. She gave me $10,000 (then a fortune) and told me that I was on my own. More than fifty years later my grandson Walter was assigned to write a high school essay about Ellis Island. "Great," he said, "I will get some original material from my grandfather." He was deeply disappointed when he learned that I had spent my first night in America in the Park Avenue apartment of Hugh Cassel.

Hasselmann: And what did you do with your new fortune?

Munk: I bought a DeSoto Phaeton and drove to California. I had fallen in love with Pasadena and San Marino Spanish street names, in such contrast with the digital New York layout. I appeared at the doorstep of the Caltech dean of admissions and said, "I am going to be a student here next year." When he said, "Let me pull your files," I had to reply, "There are no files." He was so appalled at my naiveté that he gave me a month to take the entrance exam. I holed up in a room at the corner of Lake and California streets, and passed.

2.2 Caltech

Hasselmann: How did you become fascinated with geology? Was there somebody who inspired you?

Munk: Yes, Peter Buwalda, professor of geology. At the start of classes he had casually agreed to take the class on a field trip in the event of a major earthquake along the San Andreas Fault. And there was! I remember camping out in the desert in a moonlight night and looking up at the snow covered Sierra Nevada, with Buwalda speaking about fault dynamics and the formation of mountains. I had become a good student.

von Storch: But did you have a background education in mathematics or so?

Munk: Not really, not until I came to Caltech. At Columbia night school I had taken a course in analytic geometry and freshman calculus. But here I took analysis from Harry Bateman, physics from William Houston, geophysics from Beno Gutenberg, Charles Richter, and Hugo Benioff. Caltech was very good to me.

von Storch: But no oceanography. How did you wander into oceanography?

Munk: During my junior year I was dating "Bumps" Anderson, a girl from Scripps College, who was to spend her summer with her grandparents in La Jolla. I needed

Fig. 2.9 Walter first came to Scripps for the summer of 1939 after completing his Junior year at Cal Tech. He lived at the Community House (above), now the site of the Judith and Walter Munk Laboratory of the Cecil and Ida Green Institute of Geophysics and Planetary Physics (IGPP). At the time, there were no houses between Scripps and the village of La Jolla

Fig. 2.10 Walter Munk's backview in a rowboat (1940). AIP Emilio Segre Visual Archives

a job to date her (by then the $10,000 had been spent; I'd better not talk about this). La Jolla consisted of a set of summer cottages, and the Scripps Institution. In early 1939 I drove to San Diego and asked the Director Harald Sverdrup for a summer job.

Fig. 2.11 Walter with Harald Sverdrup, Director of Scripps, in George Scripps Hall (1940). Harald was working on his epic *The Oceans*. He was a wonderful teacher; among other things he taught Walter how to write English (Harald was Norwegian). He and his wife, Gudrun, remained Walter's life-long friends. At the time, the total Scripps staff consisted of 15 employees, including the gardener. For several years Walter was the only Scripps student

I was appointed a student assistant at $50 per month. I have never lived better; for food we could pick abalones from old Scripps pier, and the chemical oceanography division provided (unwilling) assistance for parties in the old Community House (the present site of IGPP) (Fig. 2.9). I believe the total Scripps staff consisted of 18 people, including the gardener. I played tennis with the Director's beautiful Norwegian wife Gudrun, and whenever she won I would be invited to a dinner of "fiske pudding." I met Roger Revelle who had just gotten his PhD; Roger and Ellen were to be my lifelong friends. Certainly Harald and Roger had a primary influence on my entire career. By the end of the summer I had decided to become an oceanographer (Fig. 2.10).

The *RV E.W. Scripps* (a sailing schooner) had just returned from an expedition into the Gulf of California. Harald Sverdrup gave me some data and said, "See what you can do with it." It eventually became my first paper [1]: Internal Waves in the Gulf of California (not very good).

I was back next summer. My romance with Bumps had cooled, but my love affair with Scripps had begun, and it is today as it was seventy years ago. I asked Sverdrup whether he would take me on as a Ph.D. student (Fig. 2.11). He thought about it silently for an infinite thirty seconds and then said, "Okay, but I cannot think of a single job in oceanography opening over the next ten years." I quickly replied, "I'll take it."

This was 1939, Austria had been invaded by Hitler, the Schuschnig Government (including my stepfather) had fallen. My mother had gone to England; she had been a student at Cambridge University just before World War I (reading botany with Harold Jeffreys when he was still a botanist). I thought war with Germany was imminent, and enlisted with the U.S. Army. I had become a citizen in June 1939. I had applied the previous year but had flunked the test (Klaus and Hans, I must be the only person you know who flunked this exam.) In my first year at Caltech I had taken a course on the U.S. Constitution from Professor William Bennett Munro; contrary to expectation I found it fascinating. So when in my 1938 oral exam I was asked, "What is the Constitution of the United States" I happily slid into a lengthy sermon. The immigration officer cut me off, "Go study the book." The proper answer was a verbatim "The Constitution of the United States is the supreme law of the land."

2.3 Serving in the Army

Hasselmann: Tell us about your military career.

Munk: Very undistinguished. The United States did not go to war until Pearl Harbor in December 1941. Service in peacetime can be boring. I was made a Scout Corporal in the 146 Field Artillery at Fort Lewis, Washington, even though I protested that I had a poor sense of direction. The second time I lost my way I was broken to Private First Class. But there was one break: a month on Mount Rainier with the fledgling Ski Troops (Fig. 2.12). The unit later moved to Camp Hale, Colorado, with distinguished service in Europe.

In the meantime a small Scripps Group had been formed under Sverdrup and Revelle to work on problems of Anti-Submarine Warfare (ASW). The Allies were taking a heavy toll of ships being sunk by German submarines. The ASW effort was being conducted under the auspices of the University of California Division of War Research (UCDWR) with offices at the U.S. Navy Radio and Sound Laboratory at Point Loma. Sverdrup and Revelle wrote a letter to the War Department asking that I be permitted to join their group (Fig. 2.13). I jumped at the chance; my enlistment term was almost up, and I was bored listening to my first Sergeant's lectures on when to salute (and when not to).

I was discharged in December 1941, a week before Pearl Harbor, and reported for duty at Point Loma. A week later all discharges had been cancelled. My unit in the 146 Field Artillery was shipped to fight the Japanese in the Owen Stanley Mountains of New Guinea and was nearly wiped out.

Fig. 2.12 Walter (*center*) with Ski Troops on Mount Rainier, Washington (1940). After completing his master's thesis in Oceanography, Walter enlisted in the Field Artillery as a Private expecting to see early action in Europe. Two years later, he was discharged and joined the newly formed U.S. Navy Radio and Sound Laboratory in San Diego to work on Anti-Submarine Warfare under Harald Sverdrup. One week after Walter's discharge, Pearl Harbor was attacked; his Battery was shipped to the Pacific and suffered very heavy casualties

2.4 Clearance Problems

Hasselmann: That was before you began with the ocean wave prediction.

Munk: Yes. What then happened is the black hole of my career. Starting in 1941, Sverdrup, Francis Shepard, Johnson and Eugene La Fond would commute daily from Scripps to the Navy Point Loma Laboratory. One day in March 1942 Sverdrup was refused entrance; his clearance had been suspended. A month later my clearance was cancelled. It was to be fifty years before we learned what had happened.

I have written a complete account on Harald U. Sverdrup and the War Years [243] on the basis of the FBI (Federal Bureau of Investigation), Army and Navy Intelligence files requested by my wife Judith in 1995 under the Freedom of Information Act. What happened is that two Scripps professors and one Scripps technician had reported to the FBI that Sverdrup had pro-Nazi sympathies.[2] Nothing could be fur-

[2] Sverdrup had written his doctoral dissertation under the guidance of V. Bjerknes while in residence in Germany. At the beginning of World War II the German *Meteor* Expedition (with Defant's participation) was still at the forefront of sea-going oceanography. By the end of the war the leadership had passed to the United States.

Fig. 2.13 Cartoon of Walter lecturing on Anti-Submarine Warfare (ASW) (1945)

ther from the truth. As for me, it was partly because I was Sverdrup's student, partly because (according to the files) I frequently brought girls to my office at night and these girls spoke with a German accent (this must have been Sverdrup's daughter Anna). In another interview my Washington landlady testified that I was loyal but that she would never rent a room to me again because I was so untidy. Sverdrup and I did not know that for the month we worked in Washington on the North-African Landings we had been under 24 hour surveillance.

My clearance was eventually re-established, and for the last sixty years I have had broad access to sensitive information as part of my JASON activity (more later). Sverdrup returned to Norway and died without ever learning what had happened.

von Storch: You suddenly were out of business from one day to the other, right?

Munk: Yes, it was a very difficult time for Harald Sverdrup. He had lost his brother in a British Commando raid on Spitzbergen. The Nazis jailed his two sisters in Norway. He had a tremendous emotional involvement in the war and yet was not permitted to use his unique talent to contribute to the war effort. And so he returned to his beloved Norway.

von Storch: But they needed him; he was still director of Scripps.

Munk: He was still director of Scripps, and they did permit him to work on problems that did not require he look at the raw data. But such a restriction was totally unacceptable. Sverdrup's career was based on first hand contact with observations. Sverdrup's family had become American citizens, and in a letter to U.C. President Sproul he had expressed his intention to remain at Scripps after the war. I think that without the clearance problems he would have stayed for another term. He was a wonderful director. He had an enormous influence on my life.

von Storch: After the war...

Munk: ...after the war. He never knew what had happened.

Hasselmann: So he had already died by the time you received the information.

Munk: Yes, he never knew what happened. But there was an exoneration of a kind. In June 1943 the Navy Bureau of Ships issued a statement which "... expresses confidence in Sverdrup," adding that, "The Bureau is familiar with the substance of the derogatory reports... and believes that these reports are without foundation." But they did not reinstate his clearance. In my case there had been a statement one month earlier, "War Department investigation of Munk concludes that he had no pro-Nazi sympathies, and that any evidence that he was pro-German was entirely hearsay." I have previously mentioned that I have since enjoyed unusually complete access to classified information. For the last 25 years, far longer than anyone else, I have held a Secretary of the Navy Chair in Oceanography.

Sverdrup announced in January 1947 that he would be returning to Norway, and recommended Revelle as his successor. But the last few years of the Sverdrup directorship were anything but a holding exercise. Far from it. The important development was the collapse of the California Sardine Fisheries and the establishment of the California Cooperative Oceanic Fisheries Investigation (CalCOFI) modeled on the International Council for the Exploration of the Sea (ICES) in the North Sea. CalCOFI provided the ships and money for regular cruises to build oceanographic time series. Roger's experience and taste for sea-going work made him the obvious candidate for Scripps Director. But the majority of the Scripps faculty protested and the Revelle appointment was delayed until 1951, with Carl Eckart assuming the Directorship in what I would call a holding maneuver. The opposition was based on some of Roger's personal habits: late to meetings, slow in answering mail, etc. Among the leaders of the opposition were Professors Denis Fox and Claude ZoBell, the same who had blackballed Sverdrup (I have not previously mentioned their names).

Chapter 3
Bikini (1946) and Eniwetok (1951)

Hasselmann: Your clearance paved the way for your participation in the American Atomic Bomb Tests?

Munk: Yes, the 20-kiloton fission bombs in Bikini in 1946, and the 17-megaton fusion bomb in 1951 (called Ivy Mike). Let me talk about Bikini first. William Van Arx of Woods Hole and I were tasked to estimate the rate at which radioactive contamination would be flushed from Bikini Lagoon [13]. When viewed on Pacific maps, the lagoon appeared as an insignificant speck, but it was not so small when we got there. We were given 10 days to do our job. We requisitioned a Navy reconnaissance plane and rigged up a simple bombsite. Van Arx was navigator and I was bombardier. We dropped dye markers filled with a highly concentrated mixture of green hexafluoride (used to locate downed fliers) into the lagoon openings, and the colored spots were photographed over the subsequent half tidal cycle. These spots gave a rough idea of the in and outflow. There are about ten lagoon openings, and by the end of the week we had monitored nine, each showing a net inflow! The tenth (and last) channel came to our rescue, with a large net outflow. (The night before we had despaired as to how to report a violation to the principle of mass conservation.) In spite of great care, some tiny volume of the green dye would rub into my trousers. Our bunks were on *USS (United States Ship) Allen M. Sumner* (DD-692). After a few days, I noticed that the uniforms of the hundred or so officers and crew had taken on a greenish tinge. On the last day, Captain Ciano invited Van Arx and me to his quarters for hearts-of palm hors d'oeuvres. He received us in (not so perfect) dress-whites with the words, "I don't know what's wrong with the ship's laundry…"

Bikini was also the site of the perfect oceanographic experiment. The problem was to measure the maximum height of the waves caused by the underwater explosion (Bikini Baker). One member of our team, Jeff Holter, purchased a case of beer, emptied the contents, and then nailed the empty beer cans on a nearby palm tree. Following the test, the lower cans were found filled with lagoon water, the upper ones were empty, with the boundary constituting a reliable estimate of the highest run up.

H. von Storch, K. Hasselmann, *Seventy Years of Exploration in Oceanography*
DOI 10.1007/978-3-642-12087-9, © Springer 2010

Fig. 3.1 (**a**) Early days of scuba diving during Operation Capricorn (1952–1953); (**b**) Walter observing a giant coral using the newly invented Aqualung

I revisited Bikini Lagoon five years later, on the way to Eniwetok to monitor the H-bomb explosion, three orders more powerful than the Bikini tests [247]. Scuba diving has just been invented, and we were learning how to use the new gear (Fig. 3.1). I dove to the bottom of the lagoon, 180 feet (my deepest dive) and looked at the eerie silhouettes of the battleships that had been sunk during the Bikini tests. Not much had changed in five years, and the Bikini natives who had been evacuated on a moments notice had not been allowed to return. We established a bottom pressure recorder to monitor the forthcoming explosion at Eniwetok 250 n. miles to the west.[1]

von Storch: Why would you want to install a wave recorder so far from the explosion site?

Munk: Roger Revelle, John Isaacs, and I had become concerned that the H-bomb would trigger a tsunami with distant outreach, and we browbeat the Atomic Energy Commission in preparing for such an event. Looking at fathograms of the steep Eniwetok Seamount shows evidence of previous underwater landslides. This is a region of very low earthquake activity, and we were concerned that the shock of a magnitude 7 earthquake (the thermo-nuclear explosion) would trigger an underwater landslide. Such landslides are good generators of tsunamis.

[1] Willard Bascom had been diagnosed with terminal cancer. He was a passionate underwater photographer and had made it very clear that he did not want his photos of marine life spoiled by the presence of ugly fellow oceanographers. Walter was standing on the lagoon bottom with a tsunami pressure gauge raised above his head (recording a Laplace transform of instrument response) when he noticed Bascom taking a most unusual photographic interest in the calibration. When Walter finally turned around, he found himself a few feet from the object of Bascom's intense interest (Fig. 3.2). The calibration was an incomplete Laplace transform.

Fig. 3.2 Willard Bascom took this picture in 1952 in Bikini Lagoon when Walter was about to be devoured by a shark

Plans were made for evacuating low areas on islands within some hundreds of miles of the zero-point at Eluklab Island. The evacuation order was to be triggered by the actual detection of a tsunami signal. Offshore depths are typically 18,000 feet, but there was an available seamount reaching to within 4,500 feet of the surface. Willard Bascom of Scripps established two moorings, with a taut piano wire leading from the anchor on top of the seamount to a buoyant raft at the surface. A differential pressure gage with peak response at tsunami frequencies was clamped to the piano wire 130 feet beneath the surface raft. The recording was on a primitive Esterline-Angus pen and ink paper tape. Passage of the tsunami wave crest would give an increased pressure signal. For anchor, Bascom had clamped together some old San Diego trolley car wheels (the first example of what was to become standard practice for the resting places of used railroad wheels).

Bascom and I tended identical moorings on the seamount, standing on 3×3 foot rafts and anxiously observing the paper tape recording (Fig. 3.3). We were separated by two miles with the Scripps vessel *RV (Research Vessel) Horizon* between and within sight. We had arranged four semaphore flag signals:

ABLE ABLE ABLE *Destructive tsunami Pacific Ocean*
BAKER BAKER BAKER *Destructive tsunami Marshall Islands*
CHARLIE CHARLIE CHARLIE *minor tsunami*
DOG DOG DOG *No tsunami*

The *Horizon* was in open contact with the flagship *USS Estes* that in turn had open communication links to the island evacuation sites. Time zero had been set for 1952 November 1 0715.000 hours Eniwetok local time. It was before dawn, cold and wet. I put on my high-density goggles. An instant heat blast signaled the explosion. At 0721 a 5-millibar air shock arrived, followed by angry rumbling. After that, nothing.

Fig. 3.3 Willard Bascom (above) and Walter observed the hydrogen bomb test, Ivy Mike, at Eniwetok Atoll from a 3 × 3 foot raft, recording bottom pressure for a possible tsunami. Four truck inner tubes were used for floatation of the plywood raft, which was anchored to the 4500 foot deep seamount by San Diego trolley car wheels. (Willard Bascom *on raft*, John Isaacs and Monk Hendrix *in the rowboat*)

Fig. 3.4 Crossing the equator aboard the RV *Horizon*, Operation Capricorn (1953). The lettering on the researchers' chests reads RV HORIZON SIO LA JOLLA CAL UCLA. (Walter is *seated* on the *front left*)

Fig. 3.5 Homeward bound after nine months at sea on Operation Capricorn (1953). Two Scripps research vessels, RV *Horizon* and *Baird*, had participated in Ivy Mike, the H-Bomb test at Eniwetok Atoll. The return voyage was devoted to seismic and heat flow measurements which contributed vital information to the evolving theory of plate tectonics. Expedition leader Roger Revelle (*Back row, second from the left*). Walter (*Front row, second from the right*) is seated next to Win Horton; he later married Win's sister, Judith. (*Standing, left to right*: Dick Van Herzen, Roger Revelle, Willard Bascom, Ted Folsom, Alan Jones, Gustaf Arrhenius, Henri Rotschi, Robert Livingston, Russell Raitt. *Seated, left to right*: Phil Jackson, Dick Blumberg, Ronald Mason, Bob Dill, Art Maxwell, Winter Horton, Walter Munk, and Helen Raitt)

By then the mushroom cloud had reached 20 miles. I was 72 n. miles from Eluklab Island (which by then had evaporated) but the appearance was that I was beneath a raging inferno. I kept adding 5-minute time marks to the straight line drawn by the pressure recorder, At 0745 the *Horizon* came by the rafts to pick up Bascom and me. The Task Group Commander had ordered her to proceed at flank speed[2] (11.5 knots for the *Horizon*) on course 045T to avoid radioactive fallout. By noon we were hove to (as ordered) north of Ground Zero when it started to rain. The radiation safety officer recorded 30 milliroentgens per hour (the permissible outdoor rate was 7 mR/hr). We immediately initiated the fallout procedures, clothes were thrown overboard and the wash-down procedure was put into action. But by then, as Chief Scientist Roger Revelle put it, the *Horizon* "had lost her virginity" (Figs. 3.4 and 3.5). For the remaining twenty-six years of Scripps service she was unable to accommodate experiments involving low-level radiation counting.

At 1400 we received orders from the Task Force Commander to again proceed at flank speed, but now southward. After two hours the activity had decreased to

[2] The nautical term "flank speed" is its true maximum speed.

Fig. 3.6 Judith and Walter enjoying their wedding luncheon in Edward Evert Horton's garden in Encino, California (20 June 1953)

0.3 mR/hr. The cumulative total was well below the allowable personnel exposure total of 3 R. None of this would have happened if we had stayed with the rest of the fleet. We have never been able to reconstruct the reasoning behind the initial evacuation order.

Next morning we returned to the seamount to recover our gear. I un-spooled the paper tape back to 0745 the previous day when I had been ordered to abandoned the raft. Within 90 s following my final time mark, the record showed a large positive pressure jump (perhaps the pressure gauge had slipped down the mooring wire). On hindsight the signal was not creditable: too late, too large, too step-like. And there was no signal at the neighboring raft tended by Bascom. Klaus, you may remember that on my 65th birthday, I spoke of the occasion. I then thought I would have gone through with the ABLE ABLE ABLE signal. This would have set in motion the evacuation of several thousand people. I would have been too embarrassed to return to Scripps, and would have left the ship at the next landfall in Tongatapu.

Hasselmann: ... Lets not contemplate this... an early demise to your career...

Munk: My entire life would have been different...

Hasselmann: ... and what did in fact happen after your left Eniwetok?

Munk: We had made enough money participating in the bomb test to spend the next half-year doing geophysics on the way home. Russell Raitt did some seismic work and found a surprisingly thin sedimentary layer, consistent with only 100 million years of sedimentation. Arthur Maxwell found a normal heat flow through the sea bottom. This eventually became evidence for Plate Tectonics (we missed the proper interpretation). We spent a wonderful Christmas in the Tonga Islands. When we eventually returned to San Diego we had been gone the better part of a year. *Horizon* was home, and land creatures were strange and somewhat frightening. We left the vessel making suitably rude remarks to our shipmates, "Well I am glad I won't have to see you for breakfast," and to the cook "I won't have to eat your chow any more." But that evening half the contingent were back for dinner on the *Horizon*.

The long time at sea had given me time to reflect and I had decided to terminate a failing marriage. I took residence in Reno, Nevada, and filed for divorce from Martha Chapin Munk. I proposed to Judith Horton (Fig. 3.6). Judith had come down with polio on her 21st birthday, the day she was to enter the Harvard School of Design. She spent some months in an iron lung, and then came to stay with her grandmother Mrs. Oscar Kendall in Pacific Beach. By then she had recovered sufficiently to walk with a cane. But old polio victims don't get better with age. Ten years later she required Canadian Walking Sticks. It took twenty-five years for her to sit down in a wheel chair. She was a perfect partner in all I did for 53 years. Polio was a challenge, not a handicap.

Chapter 4

Settling Down at Scripps

4.1 Munk Finally Gets His Degree

Hasselmann: I would assume by then you had gotten a degree and been appointed to a Scripps faculty position?

Munk: Not yet. It was far more exciting going to Bikini than meeting the University's Ph.D. requirements. One morning Harald Sverdrup said, "If you don't get your thesis in I will have to ask you to leave." I then went into high gear and submitted a 19 page thesis in one month [6]. For that purpose I combined some work I had done on two quite different processes that lead to an increase in wave period: selective attenuation (short waves are damped more quickly) and dispersion (the interval between the first and second crest of a tsunami increases with time). Klaus, you know Jules Charney. He and I had our final examination at UCLA together, he from 2 to 3 pm, I from 3 to 4 pm. We had identical committees, including Sverdrup and Jack Bjerknes. But there was a big difference, Jules wrote a seminal thesis, mine was awful. The two processes of wave period increase have nothing to do with each other and discussing them jointly just leads to confusion [15]. In 1979 when the Dean of UCLA (where Scripps degrees were awarded in the forties) called to offer the Distinguished Alumnus Award, I thought for a moment he was going to cancel my degree. Fortunately the University of California has no mechanism for withdrawing a degree once granted.

Hasselmann: Oh, I had a Ph.D. student who went through because I missed his elementary error. In such cases one should sack the advisor, not the student. So I should have been sacked.

von Storch: Now let us see... I think we are more or less through with ocean waves. What are the next things you looked at?

DOI 10.1007/978-3-642-12087-9, © Springer 2010

4.2 Wind-Driven Ocean Circulation

Munk: In 1948 I took a sabbatical in Oslo, Norway. Harald Sverdrup had been gone from Scripps and I missed him. I had worked next to Harald's office when he discovered the "Sverdrup relation" $\beta My = \text{curl}\,\tau$ between the poleward water transport M_y and the wind curl, where $\beta = \mathrm{d}f/\mathrm{d}y$ is the poleward derivative of the coriolis parameter f. He discovered this relation not by theory, but, as was his custom, by looking at observations. Traditional geostrophy fails at the equator where $f = 0$. Sverdrup delayed the publication for months. How could such a simple result have been overlooked? He finally published in 1947.

A year later Henry Stommel published his beautiful demonstration that western boundary currents are associated with β rather than f. On my sabbatical I discovered that Stommel's equations reduce to the Sverdrup relation away from boundaries. This permitted an evaluation of the Gulf Stream transport from an integration of wind stress across the Atlantic [20]. The results, 36 Sverdrups, were within a factor of 2 of existing estimates based on hydrographic data. In the publication I used the words "ocean gyres" which have stuck ever since.

von Storch: So you did that more or less on the side. How long did you work on this?

Munk: About a year.

von Storch: This was one of the fields that you touched only once.

Munk: A year later I met George Carrier who immediately wrote down a one-line solution to the Gulf Stream as a boundary solution [22].

Chapter 5

From Waves to Tides 1958–1968

Hasselmann: You previously spoke of several attempts to move away from ocean waves studies. What was the direction you were trying to take?

Munk: For a decade we made a concerted effort to explore the long period wave spectrum between swell and tides. Here were 10 octaves of unknown territory, only occasionally visited by storm tides and earthquake tsunamis, just waiting to be explored.

We had previously installed an analog wave recorder at the end of Scripps Pier, which rejected both the "high-frequency" swell and the low-frequency tides [11]. This showed prominent oscillations of 1 to 5 minute periods that were clearly related to the amplitude modulation of the incoming swell. We called the oscillations "surf beat." They are associated with the non-linear interaction of two neighboring frequencies f_1 and f_2 in generating $f_1 - f_2$. But a proper analysis had to wait fifteen years for "Bispectra of Ocean Waves" by Hasselmann, Munk, and MacDonald [83]. Bispectra had many applications, from surf beat to the deep-water wave-wave interactions required in Klaus' wave forecasting.

Hasselmann: Were not bispectra an invention of your spectral mentor John Tukey?

Munk: Yes! I remember John coming by La Jolla one day and pronouncing that I had to learn about bispectra, "I don't know for what problem bispectra will do you some good, but I know it will do you some good."

Hasselmann: Were there other discoveries of long-period waves?

Munk: Yes, a host of different phenomena, mostly controlled by topography. The centerpiece was Frank Snodgrass' application of the vibratron transducer for measuring pressure fluctuations on the sea bottom. Recording was done on shipboard. To avoid jerking the transducer (always a problem) there had to be a separate anchor with lots of slack cable between anchor and transducer.

H. von Storch, K. Hasselmann, *Seventy Years of Exploration in Oceanography*
DOI 10.1007/978-3-642-12087-9, © Springer 2010

First let me mention measurements in the inshore environment. Some harbors exhibit pronounced and undesirable oscillations that are excited by long period wave activity outside the harbor. One would think that the appropriate repair would be to narrow the harbor entrance. But just the reverse is true; narrowing the entrance increases the resonant Q and increases resonant amplification. We called this the "Harbor Paradox" [74].

Another early application was to edge waves, gravitationally trapped waves that move *parallel* to the coastline. The amplitude decreases exponentially seaward, with a center of mass at a distance $\lambda/2\pi$. For a constant bottom slope β the dispersion law is the same as for surface waves in deep water, $C^2 = g^*\lambda/2\pi$ provided we use the "reduced gravity" $g^* = g \sin\beta$. For a representative slope of 0.01, period 1000 s, the (along-shore) phase velocity is then 16 m/s and the wave length $\lambda = 16$ km. The depth at the center of gravity is 25 m. We set up an array of pressure recorders parallel to the coast at Oceanside, California where the coast is straight for 30 km [89]. Edge waves of mm amplitude were always present and easily measured. Measured and theoretical wave velocities were in remarkable agreement; so much so that to the casual reader the theoretical dispersion curve looks just like any other least-square fit to the observations rather than a confirmation of the theory.

von Storch: But, or maybe therefore, no one has paid any attention?

Munk: Quite so, but with one exception. At the time of the experiment I met Commander (Cdr.) Jones (not his real name) at a party and he wanted to know all about edge waves. Cdr. Jones was in command of a battle cruiser and became interested in how to generate an edge wave. I suggested that moving the cruiser parallel to shore at a distance of $1/2\pi$ the longshore wave length (depth of about 12 fathoms) and the resonant speed of 32 knots should do it. The experiment was carried out next morning, but without generating an observable edge wave.

von Storch: I would think having a cruiser chase down the coast at flank speed is not a smart thing to do.

Munk: It is not. Fortunately the careers of Cdr. Jones and Prof. Munk were not adversely impacted. Let me go to another experiment. With vibratron pressure recorders at about 30 m depth, we took simultaneous readings at La Jolla and on the eastern shore of San Clemente Island, 100 km seaward [78]. We found a spectral continuum, with the two records coherent and in phase for frequencies of less than 0.7 cycles per hour and out of phase for frequencies above 0.7 cph. The measurements gave clear evidence of shelf resonances. We pushed towards lower and lower frequencies. In a *tour-de-force* Sir Edward Bullard and Munk reported that they had pushed all the way to the tidal frequencies and patched the continuum across the tidal band [86], with an underlying intensity of an astoundingly low 0.1 cm^2/cpd.

It had become quite clear that the vibratron transducer was not up to the job. Frank Snodgrass had started working with a quartz crystal pressure transducer that was less sensitive to temperature changes. More important, he was experimenting with a free-fall self-contained capsule that could be left on the sea floor for months at a time, and then recalled by acoustic command. The capsule represented half our annual budget and Frank could not get himself to toss it overboard without a safety leash. The leash would get fouled in various configurations. One day the leash became accidentally undone, and things worked perfectly. Within a few years the free-fall procedure had become routine.

Chapter 6

Deep Sea Tides 1964–2000

Hasselmann: So the tidal program was the end product of a ten-year effort of measuring offshore waves of increasingly longer period?

Munk: Exactly. But the free-fall capsule was only one of three new components of the evolving program. It always pays to know the ultimate limits set by instrument noise and background noise. Having been spoiled by a high signal-to-noise ratio, the tidal community has ignored the noise background. David Cartwright and I made the caustic remark [97] that "noise-free processes do not occur except in the literature of tidal phenomena ...," a comment that did not endear us to the tidal community.

When you allow for the interstitial noise continuum it turns out that the weaker tidal harmonics that are routinely included in the traditional predictions are hopelessly contaminated; including them makes the prediction worse rather than better. From these considerations, Cartwright and I proposed the Response Method of Tidal Prediction [97].

Hasselmann: Can we return to this later? You mentioned a third consideration.

Munk: This was the time of the first meaningful solution of THE tide problem, defined as follows: given the motions of Earth, Moon and Sun, and the dimensions (bottom and sides) of the ocean basins, compute the global tides. The mathematician Chaim Pekeris,[1] a U.S. citizen in residence in Israel, had just obtained a meaningful numerical solution using a rough bathymetry of the global oceans. Prior to that, Sir George Darwin (son of Charles) and other 19th century mathematicians had evaluated the tides for ocean basins of constant depth bounded by two lines of longitude, and other simple geometries.

[1] Chaim Pekeris was born in Lithuania and became a naturalized U.S. citizen when he took his degree at MIT under C.-G. Rossby. Later he founded the Applied Mathematics Department at the Weizman Institute where he built the computer GOLEM. The computer is named for a mythical clay giant that was miraculously brought to life by a saintly rabbi to watch over Jewish citizens in 16th century Prague.

H. von Storch, K. Hasselmann, *Seventy Years of Exploration in Oceanography*
DOI 10.1007/978-3-642-12087-9, © Springer 2010

Judith, Myrl Hendershott, and I had gone to Israel to hear Pekeris speak about his achievement. Myrl had followed Pekeris calculation with a numerical solution that allowed for the tidal distortion of the solid Earth and gravitational self attraction (the so-called Love Numbers k, h), which changed the results by a factor of order 2. I remember that Pekeris was willing to talk to me about his own work but would not listen to Myrl (then only a Scripps Post Doc). There is a curious twist to this story. Pekeris had done his tidal work under an NSF (National Science Foundation) grant that allowed him to build GOLEM, one of the early electronic computers. When Judith and I called on Pekeris, we were astounded to see his premises in the Weizman Institute guarded by soldiers with loaded weapons. It was at about that time that rumors were current that Israel had developed an atomic bomb. Now the design of the bomb (as well as the solution of the tide problem) requires extensive computing facilities. Could Pekeris have helped design the atomic bomb with a computer provided by NSF under the pretense that he was working on the tide problem? This needs to be checked. Memory is not history.

Hasselmann: Numerical solutions of the tide problem (as you call it) must be one of the earliest applications of electronic computers.

Munk: I believe it was. John von Neumann, who is credited with having pioneered the design of electronic computers, was visiting La Jolla at the time. We spoke about the application to the tide problem, and he immediately asked, "What do you do about the coastal boundary condition?" A very perceptive question, which is still being worked on.

von Storch: Where did you go for support of the tide problem?

Munk: To both ONR and NSF. I remember being initially turned down with the words that, "The tide problem had gone to bed with 19th century mathematicians." I protested that we had planned for a much broader attack: to compare theoretical deep-sea tides with measured deep-sea tides, using the Response Method for analysis.

Tides along the American west coast can be represented by a rotational edge traveling northward with velocity of order $\sqrt{(gh)} \approx 200\,\mathrm{m/s}$ and long-shore wave length $\lambda_{\mathrm{rot}} = \sqrt{(gh)}/f$ = order 10,000 km and a seaward extent of order $\lambda_{\mathrm{rot}}/2\pi$.[2] (Recall that gravitationally trapped edge waves have λ of order 10 km and can go either way.) Working with Jim Irish and Mark Wimbush, we first did some deep-sea drops off California [119, 126] and located the Northeast Pacific M_2 amphidrome (zero amplitude). This was followed by drops between Australia and Antarctica, to explore the latitudes where the sublunar point travels around the southern oceans at the speed $\sqrt{(gh)}$ of free waves. A naïve theory predicts resonant amplification at

[2] $f = 2\Omega$ sin latitude is the coriolis parameter. This horrible notation is due to Carl Gustav Rossby. He developed the equations for the Rossby waves while visiting Sverdrup in La Jolla. I was present when he was on the beach writing equations in the sand, and threatened by a rising tide introduced f as an instant shorthand.

Fig. 6.1 Walter (standing in sea-going portable laboratory) discussing deep-sea tide measurements with Frank Snodgrass (*third from the left*) aboard the RV *E.W. Scripps* (circa 1973). Frank pioneered the use of sea-going portable laboratories

such latitudes. We didn't believe the theory but made measurements anyhow. The result was a rather dull transition from Australian to Antarctic tides.

We organized an international SCOR (Scientific Committee on Oceanic Research) working group WG27 to explore deep-sea tides with the free-fall drops. Numerous measurements were made, particularly by Cartwright in the U.K. and Hal Mofjeld of NOAA (National Oceanic and Atmospheric Administration), Miami. Snodgrass participated in an international calibration experiment in the Bay of Biscay. Eventually about 200 pelagic tide stations have been occupied by different investigators, and these provided a check on the numerical modeling of tides (Fig. 6.1).

Our last drops were made in 1974 south of Bermuda in 5.5 km of water, as part of the MODE (Mid-Ocean Dynamic Experiment) experiment [142]. We discovered unexpected pressure fluctuations at subtidal frequencies that are coherent over 1000 km [143]! As far as I know these have not been explained. With regard to tides, an analysis by Bernard Zetler was in splendid agreement with the traditional Atlantic cotidal charts [145]. Two independent drops in the same area gave M_2 amplitudes of 32.067 and 32.074 cm. When it comes to four-figure accuracies, it is no longer oceanography. Further, satellite altimetry looked increasingly promising for future measurements of deep-sea tides. It was time to move on.

von Storch: You previously mentioned the Response Method of Tidal Prediction. What is that all about?

Munk: Cartwright and I proposed what we thought was a significant change in the method of tide prediction [97]. I will need to write a bit of mathematics. Let $x(t)$ designate the tide producing forces, $y(t)$ the spike response and $z(t)$ the predicted tide, all referred to one particular tide station. Then the convolution integral gives the predicted tide, $z = x \cdot y$. The harmonic method consists of evaluating the station tide spectrum $Z(f)$ from a station record $z(t)$ (using capitals for Fourier transforms) and then predicting future $z(t)$ from a Fourier transform of $Z(f)$. The trouble is that $Z(f)$ is very complex, with the principal diurnal and semidiurnal lines split by monthly modulation, with further fine splitting by the annual modulation and hyper-fine splitting by the lunar 18.6 year modulation.

The discrete frequencies are not at equal intervals (as in classical harmonic analysis) but occur at $f_{ijk} = c_i cpd + c_j cpm + c_k cpy + \ldots$ where the c's are integral multipliers of the daily, monthly and yearly frequencies. Weak lines are improperly enhanced by including some of the noise continuum. There is no reference to the gravitational theory of tides (except for providing the f_{ijk} frequencies). In the response method we evaluate the tide producing forces $x(t)$ directly from the known motions of Earth, Moon and Sun in the time-domain, and then use the station record $z(t)$ to evaluate the station response $y(t)$ once and for all. It turns out that the station admittance $Y(f)$ is vastly simpler than $X(f)$; there is no need of evaluating the infinitely complex spectrum $X(f)$ or $Z(f)$. In some tests by Zetler *et al.* [123] the response method come out better (but only slightly) than the harmonic method.

Hasselmann: So you improved one of the few geophysical predictions that already work well.

Munk: Guilty. But for very short records (such as the deep-sea recordings) the improvement was substantial.

Hasselmann: How about shallow regions with strong "overtides"?

Munk: That is an important point. For very flat shelves with strong nonlinear interactions the response method can easily be extended by a formalism parallel to extending a spectrum to a bi-spectrum...

Hasselmann: I see. Tukey again to the rescue – although I guess the use of nonlinear response function expansions in the time domain probably preceded their application in the frequency domain.

Munk: Perhaps, we at any rate were happy to work in either the frequency domain or time domain, whichever was more efficient for the problem at hand. Essentially what the response method does is to keep an open mind on what side of the Fourier transform is more compact. The three-body problem Earth-Moon-Sun has an exceedingly complex spectrum and the time domain is the domain of choice; if our world were associated with two-body tides (double-star without moons) it could be the other way around, the harmonic method would be the method of choice.

von Storch: How did the oceanographic – or tidal – community respond to your emphasis in this case on the time domain instead of the time-honored frequency domain?

Munk: At one time I booked myself into an international session on tide predictions; I think it was in Brussels. After my talk, 6 seconds of resounding silence. Then, "Next paper, please."

Hasselmann: I know the feeling. Were the available computing facilities adequate for the job?

Munk: Which method is more efficient depends of course on the software tools available. That reminds me of our yearlong diversion in 1965 into writing a computer program called BOMM [91] that you mentioned at the beginning, Klaus. It was a great help in our early spectral analysis applications in the frequency domain. It was a crude forerunner to MATLAB. To compute the tide potential for any given date, it was necessary to allow for the loss of 10 days (5 October–14 October 1582) in the transition from the Julian to the Gregorian Calendar.[3] This made it possible to compute the lunar orbit in antiquity. Ancient eclipses provide important information about the slowing in the Earth's spin.

von Storch: You appear to have looked into the history of tidal prediction rather closely, going far back in antiquity. Did you ever study pre-Newtonian attempts at tide predictions?

6.1 The Alleged Suicide of Aristotle

Munk: Aristotle tried to predict the tidal currents through the strait of Euripus; there is a widespread story that when he failed he threw himself into the turbulent rapids. Adrian Gill[4] and I thought this was a dangerous precedent for oceanographers, and we decided to investigate. So our combined families converged on Chalcis on August 1981.

M_2 tides in the Mediterranean are unusually low and so the cancellation at neap tide is almost complete, leaving two days in each fortnightly cycle to be dominated by wind tides. Evidently Aristotle did a pretty good job of predicting the astronomic tides (even without the benefit of gravitational theory) but was unable to cope with the meteorological tides (we still can't). Regarding Aristotle's demise we were unable to come to a firm conclusion, even after days of spirited discussions with local historical experts in Chalcis pubs.

[3] For anyone requesting the tide potential on one of the lost days, 7 October 1582 say, a note would appear, "Any son-of-a-bitch knows that these dates are missing." We expected angry phone calls, but none ever came.

[4] Adrian's book, Gill, A.: *Atmosphere-Ocean Dynamics*. Academic Press (1982), remains my favorite on the subject.

Hasselmann: Yet another open problem! I gather from you that the subject of tides had, at one time, been considered as having been put to bed, only to come up for a rude awakening.

Munk: ... like a dharma doll. This has happened a number of times, the first time after the publication of Newton's *Principia* in 1687. This gave the "equilibrium tides," the appropriate response for an ocean with time constants very short as compared to the semi-diurnal tidal forcing. Tides of the solid Earth, with normal modes of order one-hour period, come close to equilibrium theory. But the oceans have resonant periods of order fractions of days, giving resonant responses to tidal forcing. For example, the Atlantic Ocean has a resonance near 12 hours. This is the result of the strange coincidence that the depth h and width L of the oceans are such that $L/\sqrt{(gh)}$ is of order of half a day. Ocean tides require a *dynamic* theory of tidal response, as given by Laplace in *Mécanique Céleste* in 1800. (Incidentally Laplace initiated a crude method of tide prediction which resembles the response method.) For the second time the tide problem was considered solved. I suppose the third time it was solved was after Kelvin and George Darwin developed a practical method of ocean tide analysis and prediction. I would like to think that the advent of measuring tides offshore constitutes an important chapter. Here the contributions of David Cartwright stand out; it is true that he stood on the shoulders of giants. But with regard to contributing to the understanding of ocean tides as they are observed, he is second to none.

Hasselmann: The problem of tides has always been somewhat apart from the core problems of physical oceanography. And the community of tidal workers was somewhat separate from the general oceanography community. Would you agree that this suggests that tidal processes, although important and interesting in their own right, do not play a vital role in ocean dynamics?

Munk: I agree. There is at least one important exception: ocean mixing. This has a fascinating history, highlighted by great insights and curious errors. The first indication of a departure from Newtonian orbits was given by Halley in 1695; his "modern" observation indicated that the Moon had accelerated relative to the orbits indicted by ancient eclipses by 10 arcsec/century^2. Sixty years later Emanuel Kant in a paper with a title the length of a normal abstract suggested that the lunar acceleration was consistent with tidal energy dissipation. But then Laplace in 1787 announced that he had computed a lunar acceleration of 10.18 arcsec/century^2 from planetary perturbations of the orbit, mostly Jupiter (because it is so large) and Venus (because it is so close). This was considered a major triumph of 18th century science.

In 1853 Adams found Laplace had made an error and that the correct answer was but one-half of Laplace's result. This required some additional phenomenon (such as tidal friction). But no one paid any attention, because it destroyed an acclaimed triumph. Not until G.I. Taylor's estimates of tidal dissipation in the Irish Sea, followed by Jeffrey's global extrapolation, was tidal dissipation accepted as a factor in orbital dynamics.

This produced a number of independent estimates of global dissipation. They all agreed within accepted error limits. The most precise information eventually came from lunar laser ranging using the retroflectors placed on he Moon in 1969 during the Apollo mission. The semimajor axis of the Moon's orbit is increasing at a rate of 3.8 cm/year, yielding a dissipation of 2400 GW from the M_2 tides.

My interest was aroused by an early study of what we now call the Meridional Overturning Circulation (MOC). The formation of bottom water in the winter, mostly in the North Atlantic, of 25 Sverdrups would fill the ocean basins with dense, cold water in 3000 years. But this does not happen. The simplest model is one where vertical upwelling of cold water is balanced by downward diffusion from the warm surface layers. This requires energy. Some very rough calculations gave 2000 GW, with very large error limits [231].

The similarity of the two numbers hits the eye; could the tidal dissipation provide the energy for pelagic mixing? There are of course many difficulties. Perhaps the outstanding difficulty is that only a fraction of the 2400 GW is available for pelagic mixing; some, perhaps most, is dissipated in shallow seas, and Harold Jeffreys in his 1920 paper "Tidal Friction in Shallow Seas" claimed all of it.[5] In 1968 I gave the Harold Jeffreys Lecture [112] "Once Again – Tidal Friction" with the opening sentence, "In 1920 it appeared that Jeffreys has solved the problem of tidal friction. We have gone backwards ever since." Thirty years later I returned to the subject [229] under the title, "Once Again: Once Again – Tidal Friction."

In 1997 Carl Wunsch and I summed up the evidence in a "Child's Guide" for tidal mixing entitled "The Moon, of Course" [230]. By then reaction to the proposal that the Moon played a significant role in deep ocean mixing had taken a sharp turn from being considered "lunatic" to being "well known;" Carl and I preferred the lunatic era.

Hasselmann: This is an interesting question: would we have a completely different global ocean circulation if there were no tides? My impression is that ocean circulation modelers are still quite happy to ignore the tides and consider only wind and radiation driving forces, together with various empirical mixing-type diffusion coefficients and bottom friction. It would be an interesting experiment to test your concepts in a global ocean circulation model. I believe you suggested that the tidal dissipation is a two-step process. First a scattering of surface tides into internal tides by bottom topographic features – beautifully visible nowadays on satellite images – and second the conversion of the internal tidal energy into small scale turbulence. But that brings us to the next significant topic of your interest, internal waves. Tell us something about the history of internal waves as you see it, and your personal involvement.

[5] Jeffreys, H.: Tidal Friction in Shallow Seas. *Philosophical Transactions of the Royal Society of London*. Series A, Containing Papers of a Mathematical or Physical Character. 221, 239–264 (1920)

Chapter 7

Internal Waves 1971–1981

Munk: Scandinavian Fjords in late summer have a thin layer of fresh melt water above the salt water. A moving ship with its keel extending into the lower layer generates an internal wave at the fresh/salt water boundary (in addition to the surface wave wake). This greatly increases the wave making resistance. Vikings have pictures of underwater sea monsters hanging on to their boats.

von Storch: I did not know about this folklore ... but when were internal waves actually discovered?

Munk: The theory goes back to G.G. Stokes in 1847. According to the *Ocean Bible* by Sverdrup, Johnson and Fleming, the earliest measurements were by Bjorn Helland-Hansen and Fridtjof Nansen in the Norwegian Sea in 1909. Sverdrup told me that his account of internal waves led to the only criticism of the *Ocean Bible*. Hans Pettersson complained bitterly that his father Otto Pettersson's earlier measurements in the Kattegat had been neglected. Otto Pettersson had discovered internal tides breaking over the bank that separates Gullmarfjord from the sea and spent much of his subsequent career trying to convince his colleagues that tidal mixing is a factor in ocean climate. When Judith and I visited Hans Pettersson at his Institute in Goteborg in the 1960's he was still angry. According to him there are three ways of demolishing a paper: by claiming the conclusions to be wrong, to be obvious, or to have been published previously; "with regard to my father's work they used all three arguments."

von Storch: What was your first contact with internal waves?

Munk: In summer 1939, when I first came to La Jolla, Sverdrup assigned me the analysis of some temperature data taken earlier that year by the schooner *E.W. Scripps in* the Gulf of California. This led to my first publication [1], "Internal Waves in the Gulf of California" with the conclusion that the observed oscillation cold be conciliated with a standing internal wave of seven-day period. There is some curious resemblance to Otto Pettersson's work. If you don't mind, I would rather skip to the internal wave work some thirty years later.

H. von Storch, K. Hasselmann, *Seventy Years of Exploration in Oceanography*
DOI 10.1007/978-3-642-12087-9, © Springer 2010

von Storch: All right. You are now referring to the widely quoted GM Internal Wave spectrum?

Munk: Yes. Chris Garrett and I had decided in our 1972 paper [130] to allow for built-in obsolescence by calling it "GM." To our amazement it is still, as we speak, being referred to as some kind of a standard.

von Storch: To what do you attribute the longevity?

Munk: The need for some kind of standard for inter-comparison of different data sets. Chris Garrett arrived in 1970, a new product of the famed DAMTP (Department of Applied Mathematics and Theoretical Physics) in Cambridge, England. He had declined a post-doc (few ever did) because he wanted to be closer to observations. I was reminded of young Sverdrup when he declined an appointment at the Bergen School to get his feet wet on the *Maud* expedition in the Arctic Ocean.

Chris and I started looking at what was by then a very voluminous literature on temperature, salinity, and velocity as functions of time, horizontal distance (leading to towed spectra), and depth (dropped spectra). For simplicity we chose a spectrum that could be factored into a function of frequency times a function of vertical wave number and took a cavalier attitude towards boundary conditions (Rosenbluth called them the Tijuana boundary conditions: topless and bottomless). To our delight the great majority of the diverse measurements taken at different times and places in the open global oceans was consistent within a factor of two with a simple model spectrum. This was a far cry from the original notion of internal waves as an exotic phenomenology.

Hasselmann: This is not the last time that a commonly occurring process was considered to be of rare and distinct occurrence. Think of mesoscale variability.

von Storch: But, in your view, is the GM spectrum still in good standing?

Munk: No. But it was to be thirty years until Rob Pinkel showed that arctic observations were inconsistent with the assumed factoring of the spectrum. By then Chris had gotten nervous and claimed that the G in GM referred to his great uncle Arthur Garrett. A few years later Pinkel demonstrated that one could go a long way with just two Doppler-smeared spectral lines: the M_2 tidal frequency and the local inertial frequency. Here I refer to the smearing of the spatial finestructure by the vertical orbital motion of the long internal waves. Curiously enough Chris and my first joint paper dealt with this very interaction [124]. Thirty years later Chris was given a 65th birthday party (Fig. 7.1), and I had been instructed to award him the William Leighton Jordan Esq. Award.[1] Instead I chose to present the award to the GM spectrum [261b].

[1] The award is an invention of Henry Stommel, to be "given annually to the oceanographer who makes the most misleading contribution to his field. Ignorance and utter incompetence do not automatically qualify."

Fig. 7.1 Christopher Garrett and Walter demonstrating the appropriate use of Parker MacCready's scale at the Garrett 65th Birthday Symposium in Victoria, British Columbia, Canada (2008).

von Storch: How active is the subject of internal waves being pursued today?

Munk: There has been a renaissance brought about by the fact that internal waves can be seen on satellite altimetry. One thinks of internal waves of having large internal vertical displacements and negligible surface displacement (unlike surface waves). But negligible is not zero. Using satellite altimetry, Gary Egbert and Richard Ray have traced internal waves of tidal frequency from their origin over the Hawaiian Chain all the way to the Aleutians.

It follows that standard tide gauge records have a small contribution from internal tides. These are sensitive to changes in ocean stratification. The Honolulu tide gauge goes back to imperial days. John Colosi and I [257] have attributed an increase from 16.1 to 16.9 cm between 1915 and 2000 in the Honolulu amplitude of M_2 to a change in phase of the internal tide component.

Hasselmann: Actually, I have the impression that the puzzling ubiquity of the GM spectrum has triggered innumerable theoretical and experimental investigations not only in the past but even today – remember, for example, the excellent presentation on the distribution of internal wave energy in the Pacific Ocean by Jennifer MacKinnon at your 90th birthday symposium. What is the basic dynamics responsible for the universal GM spectral form? For example, while I was at the Woods

Hole Oceanographic Institution (WHOI) in 1971–1972, WHOI implemented a sophisticated tripod-mooring array of current meters and thermistors to measure the detailed modal structure of the fluctuations in the GM band. The resultant IWEX (Internal Wave Experiment) spectrum[2] *largely supported the GM internal wave model, but non-vorticity-conserving shear currents were also found to contribute to the variability. The universal form of the GM spectrum has been attributed by Müller and Olbers (1975)*[3] *to the redistribution of the energy input (from the wind or topographic interactions with barotropic tides) via resonant wave-wave interactions – in analogy with the universal spectral form of wind-generated surface waves. And a number of modeling and data-assimilation exercises are currently in progress to test the impact of competing hypotheses on the origin of vertical mixing in the oceans on the global ocean circulation. So the publication of the GM spectrum has indeed been extremely fruitful for oceanography, both in the past and still today.*

[2] Müller, P., Olbers, D.J., Willebrand, J.: The IWEX spectrum, *J. Geophys. Res.* **83**: 479–500 (1978)

[3] Müller, P., Olbers, D.J.: On the dynamics of internal waves in the deep ocean. *J. Geophys. Res.* **80**: 3848–3860 (1975)

Chapter 8
Ocean Acoustics 1974–Present

Hasselmann: In 1979 you and Carl Wunsch wrote a paper [157] *entitled "Ocean acoustic tomography: a scheme for large-scale monitoring." By then you had worked in oceanography for more than thirty years without, to my knowledge, being involved in ocean acoustics. This had largely been the domain of oceanographic specialists involved in Anti-Submarine Warfare (ASW). What made you go into ocean acoustics?*

Munk: The mesoscale revolution. This called for a radically new sampling strategy; a few ships chasing independently across the oceans at 10 knots were not up to it. Carl and I thought that a method based on acoustic transmissions at 3000 knots could work.

von Storch: So once again you entertained the community by inventing new terminology, this time "Acoustic Tomography."

Munk: Yes. We deliberately chose a name that would make people sit up and want to find out what we were talking about. CAT scans (for Computed Axial Tomography), as you know, are used by the medical profession in a related way. One measures the attenuation of electromagnetic radiation through a man's skull along many, many different paths. From these measurements one then reconstructs what is inside the skull. Here we used traveltime instead of attenuation, but the principles are the same. There is a theory that for an infinite number of "path integrals" the interior function can be determined to infinite precisions. For a given configuration, inverse theory provides the error bars. Carl had pioneered the application of Inverse Theory to oceanographic explorations, something that had been sorely lacking.

Hasselmann: So you could pull together a set of new tools, just as you did when you went into the exploration of tides.

Munk: Exactly! Let me list some of them.

1. Inverse Theory, which *ab initio* provides the variance of each estimate.
2. Perhaps the outstanding feature of long-range acoustic transmissions is the great variability from one transmission to the next. We needed a model for the under-

H. von Storch, K. Hasselmann, *Seventy Years of Exploration in Oceanography*
DOI 10.1007/978-3-642-12087-9, © Springer 2010

lying temperature (actually soundspeed) noise. Earlier Russian work had taken a Gaussian model of isotropic, homogeneous variability. But the ocean is neither isotropic nor homogeneous. The GM model, though imperfect, provided an adequate model of the underlying small-scale variability.

3. We found it useful to define two idealized models of vertical soundspeed profiles: the temperate and the polar profile. They follow from a designation of the buoyancy (Väisälä) frequency according to $N = N_0 e^{-z/h}$ and $N = 0$ respectively. I endowed the temperate profile with additional authority by naming it the "canonical sound channel."

von Storch: Another one of your interesting new terms.

Munk: Guilty, again.

4. Previous long-range transmissions, like the 406 Hz 1250 km Eleuthera to Bermuda transmission were CW; significant advantages are gained by using broadband signals permitting pulse-compression. We learned about m-sequences from Ted Birdsall.

von Storch: m-sequences?

Munk: This is a sequence of zeroes and ones that have a uniquely delta-function like auto-covariance. By correlating the received record with a stored replica of the transmitted sequence you get a superior measure of travel time, and hence of sound speed.

von Storch: You have listed four considerations for selecting ocean acoustic tomography. Are there others?

Munk: 5. We were permitted, with certain restrictions, to use the Sound Surveillance System (SOSUS) receiver arrays installed by the Navy during WWII.

There are many other innovations. It has taken twenty years to pull all this together. Among the people that made it work are Robert Spindel, Peter Worcester, Ted Birdsall, Kurt Metzger, Bruce Cornuelle, Matthew Dzieciuch, Bruce Howe, John Spiesberger, and Mike Brown. But the incentive clearly came from what I call the "mesoscale revolution."

von Storch: And what do you mean by "mesoscale revolution"?

Munk: For centuries oceanographers had followed the model of a time-invariant ocean circulation. Changes were attributed to changes in position rather than changes in time. An estimate of 10 ± 1 cm/s was considered more representative of ocean currents than 1 ± 10 cm/s. We now know that the reverse is true, and that over 95% of the kinetic energy is associated with mesoscale variability (ocean weather). This faulty concept of time invariability was upheld by an oceanographic tradition of never occupying a station twice. In the rare cases of a repeat station any change could always be attributed to instrument malfunctioning.

It had been known since the time of Benjamin Franklin that there was something wrong with that picture. Fridtjof Nansen knew there was something wrong. But it took a non-traditional oceanographer like Fritz Fuglister to really break with tradition.

Woods Hole Director Columbus Iselin was uncomfortable with the background of the people who populated the oceanographic profession. At one time there had been a strong component of people who owned yachts or knew people who owned yachts (like Columbus); but I think he was also suspicious of the university people who leaned too heavily on their ability to manipulate differential equations. Among the Woods Hole faculty were Alan Woodcock and Fritz Fuglister. Woodcock had been a sailor on the original *RV Atlantis* when she first sailed in Woods Hole under Captain Iselin; Fuglister has been a starving Falmouth artist. And they certainly have left their mark!

8.1 The Gulf Stream Sheds Eddies

Munk: Gulf Stream eddies are the most dramatic manifestation of what is wrong with a steady ocean. Fritz had become the biographer of the Gulf Stream and had a sixth sense about its behavior. What was needed here was to have a number of ships occupy stations again and again. And Fritz talked the Navy into giving him six ships to undertake the Multiple Gulf Stream Experiment.

Scripps had been invited to participate and I signed up. Fritz invited me to join him on the Command Ship, the *USS San Pablo*. We had stumbled on a lucky moment: a Gulf Stream Ring about to be shed. Two oppositely traveling branches of the Gulf Stream were within 10 miles of one another. Fritz had us cut across the two branches, reversing course every two hours. The Captain had left orders to wake him with every change. Of course; he didn't get much sleep.

von Storch: Can you tell a little what people thought about eddies at that time. Did they think the ocean was an eddy free environment?

Munk: No, they knew there were eddies. And they knew the Gulf Stream was not steady. But all this had to be put together.

von Storch: And then what happened?

Munk: This was just the beginning of a number of experiments. I want to mention in particular the subsequent Swallow Float measurements. John Swallow, a British oceanographer, had designed a float that could be ballasted to drift at a given depth. Its position could be accurately followed from a surface vessel using an acoustic signal. (This was the forerunner of acoustic oceanography.)

Stommel in his work on the general circulation had expected to find a deep undercurrent beneath the Gulf Stream, going with low velocity in the opposite direction to the surface current. I believe the early Swallow Float experiment was designed to test this model. The result put everybody on his feet. As I remember, the measured currents were ten times those expected. But to add insult to injury, two floats at the same depth and only 10 km apart (and expected to perform in parallel) were moving in opposite directions.

Hasselmann: I wonder if the reason that the steady-state picture persisted for so long was that it was inferred largely from the density structure? This is equivalent to viewing the ocean with a low-pass filter. And the overall picture of the mean transports inferred from the density structure appeared reasonably consistent.

Munk: That may well be the explanation.

von Storch: I also think it may have to do with a certain intellectual laziness. It is much easier to think of a stationary ocean circulation with just a few tidal periods superimposed. And then what happened?

8.2 The MODE Experiments

Munk: This led to the Mid-Ocean Dynamic Experiment (MODE).

von Storch: So MODE was a Woods Hole initiative. But Scripps was also involved?

Munk: Lots of groups became involved, including Scripps. I attended a fascinating planning session in Bermuda. It was there that Stommel and A.R. Robinson developed initial strategy using an old outdoor blackboard that no one could read.

MODE permanently changed the face of oceanography. Our response was the development of Ocean Acoustic Tomography.

8.3 Ocean Acoustic Tomography

Munk: Among the forerunners were the Garrett–Munk (GM) collaboration on internal waves starting 1971 [125, 139], the Zachariasen–Munk estimates of acoustic scintillations through a GM internal wave field [148], followed by the 1976 measurements by Peter Worcester and Frank Snodgrass of oppositely-directed transmissions between two deep "transceivers" 25 km apart.[1] Variations in the *average* of

[1] Worcester, P.F.: Reciprocal acoustic transmission in a midocean environment: Fluctuations. *J. Acoust. Soc. Am.* **66**: 1173–1181 (1979), Worcester, P.F.: Reciprocal acoustic transmission in a mid-ocean environment. *J. Acoust. Soc. Am.* **62**: 895–905 (1977)

the opposite travel times told us something about the soundspeed (hence temperature) profile in the intervening ocean, *differences* in travel time (with and against the current components) gave information about water movement. This is a powerful technique for measuring ocean features on a scale of tens of km, and it whetted our appetite for acoustic monitoring of the intense mesoscale features with typical dimensions of 100 km. For a 1000 km transmission of one-week duration we estimated the "noise" associated with a GM75 internal wave spectrum at 20 ms. The expected mesoscale signature was many times that large!

We formed a joint venture involving Robert Spindel of Woods Hole, Ted Birdsall of Michigan, Carl Wunsch of MIT (Massachusetts Institute of Technology) and our group at Scripps. In October 1978, Spindel put out a 2000 m deep mooring south of Bermuda. Our graduate student John Spiesberger monitored a coastal station 1000 km distant. The formation of the group was somewhat of a shotgun wedding under the persuasion of Hugo Bezdeck of ONR (and like other shotgun weddings has been one of outstanding stability). Results were promising: about a dozen distinct arrivals that could be clearly traced over the two-month transmission period. We were ready to propose a tomography array.

Hasselmann: I seem to remember that your original proposal was declined.

Munk: Yes. A proposal had been submitted to NSF to augment the existing ONR support. One of the reviewers wrote, "Travel times along ray paths are meaningless in a saturated environment." He continued that individual ray arrivals could not be resolved, and even if resolved could not be identified, and even if identified would not be stable. The proposal was declined. Our 1978 Bermuda transmissions came just in time to save us from an early demise. We responded by sending a "dot-plot" of the 1978 transmissions with a single sentence, "We have resolved, identified and tracked 13 rays for 2 months, see figure." The proposal was accepted.

Hasselmann: And why was that anonymous reviewer so far off the mark?

Munk: He was not. It has taken us thirty years to realize how lucky we were. Scattering can certainly destroy rays and all the good things that come with geometric optics. It is only now [256] that Dan Rudnick could demonstrate, using thousands of Monte Carlo transmissions, that in a flat ocean with refracted rays of small inclination, scattering occurs mostly *along* rays, and the ray structure is not destroyed. Further, the eastern Atlantic has turned out to be an unusually favorable environment for long-range acoustics. We were lucky.

von Storch: So you started with emphasis on the newly discovered ocean weather. I seem to remember more recent experiments focusing on ocean climate. How did this change in emphasis come about?

Munk: There has indeed been a gradual progression towards larger scales. The resemblance to our earlier discussion of ocean waves is interesting. There we started with ordinary wind waves, followed by exploring the topography controlled open

Fig. 8.1 Peter Worcester and Walter floating in the Blue Lagoon in Iceland following the Greenland Sea Expedition (1989)

sea seiches, followed by deep-sea tides. Here we went from mesoscale mapping (ocean weather maps) in the presence of internal wave noise to gyre scale mapping (ocean climate) subject to mesoscale noise.

There were a series of 300 km experiments in the 1980's, aimed at mapping mesoscale eddies. Although the number of acoustic paths between N moored transceivers increases like $N!$, there were never enough moorings to get good spatial resolution. To improve the resolution, Snodgrass and Cornuelle made a heroic effort using a moving source ship. Eventually it turned out that a good solution is to combine the good spatial resolution of satellite altimetry with the good temporal resolution of ocean acoustic tomography.

I count 21 sound-related experiments between 1976 and 1994 (Table A.1 of [225]). Some of the most interesting are not tomography related. I have previously referred to the pioneering two-way transmission experiment by Worcester and Snodgrass. This was on a 25 km range. Malanotte-Rizzoli and others used bottom-mounted instruments under the Gulf Stream. Worcester and Spindel measured the barotropic vorticity in the central North Pacific (turned out much bigger than expected from Sverdrup dynamics). Worcester, Sutton, and others installed a 200 km Pentagon array in the Greenland Sea (Fig. 8.1) in summer 1988 and retrieved it in summer 1989, observing the evolution of an overturn event in the intervening winter. Worcester monitored the in and outflow through the Straits of Gibraltar. Various measurements provided critical tests of the equations of state of seawater, and led to a correction of the Del Grosso sound speed algorithm. These and other experiments demonstrated a wide range of applications for the technique.

And we were gradually evolving towards ever larger-scale transmissions. In 1977 Spindel put out a 2000 m deep mooring south of Bermuda that was monitored by Spiesberger off the east coast 1000 km distant. By 1984 Birdsall and Spiesberger conducted some 4000 km transmissions between Hawaii and the west coast. In 1991 we carried out the Heard Island experiment for testing the feasibility of using global scale acoustic transmissions. (The next step will be more difficult.) Here we are talking about a few widely separated acoustic paths, and we called it "Ocean Acoustic Thermometry."

von Storch: There you go again, inventing new terms. The Heard Island experiment elicited perhaps a greater response, positive and negative, than anything else you have done. What do you think about it, now that eighteen years have passed?

8.4 Heard Island

Munk: I have mixed feelings. When we returned from the Indian Ocean, *Der Spiegel* carried an article entitled "Radau in der Tiefe."[2] Others referred to "The Shot that was heard around the World." The publicity led to a vastly enhanced interest in the impact of sound on marine life. With it came restrictions, some unwarranted I think, on the use of sound for ocean studies. We have never recovered.

von Storch: We will return to the marine mammal problem. How did Heard Island come about?

Munk: I was itching to test the range limits of man-made acoustic signals. In 1960 the *RV Vema* and *HMAS (Her Majesty's Australian Ship) Diamantina* had detonated three 300 lb explosives near the sound axis off Perth, Australia. The detonations were clearly recorded by axial hydrophones off Bermuda at a 178.2° range (180° is antipodal). The recording was considered remarkable by John Ewing (brother of Maurice) as much of the great circle is blocked by Kerguelen Bank. Allowing for lateral refraction and Earth flattening moved the ray path to the north of the shoals but at the expense of colliding with the African continent [190]. We now think that the intense mesoscale eddy activity off Cape of Good Hope allows scattered arrivals around the Cape. The problem is not solved.

In 1989, Andrew Forbes from the CSIRO (Commonwealth Scientific and Research Organization) laboratory in Hobart and I started planning a repeat antipodal transmission, but with non-explosive sources [194]. We figured a decrease in travel time from global warming of 0.1 to 0.2 s per year, permitting detection in a decade. For

[2] Radau in der Tiefe. Der Spiegel **32** (1991)

source location we chose Heard Island, an uninhabited Australian island in the south Indian Ocean, with direct unimpeded geodesic paths into all five ocean basins, westward past Africa into the South and North Atlantic, eastward past New Zealand into the South and North Pacific, and northward into the Indian Ocean. A southward path towards Antarctica was readily available. And by 1991 with the enthusiastic support of the Navy Oceanographer Admiral Richard Pittenger we were in Freemantle in western Australia ready to depart.

We used existing HLF-4 low-frequency sources of high intensity (217 dB re 1 μPa at 1 m), that could be lowered through the center well of the *RV Cory Chouest.* Rather late in the planning stage we had been required to obtain a NOAA permit for turning on the sources (we were using American government equipment). And soon thereafter the Australian Environment minister followed with the requirement of a permit. We had scheduled the first transmission to start at 0000 Greenwich Mean Time on 26 January, Australia Day. Oceanographers from 9 countries aboard 12 ships in all of the world's oceans (except the Arctic Ocean) were readying their equipment. Postponement would have been tantamount to cancellation. So we left on our two weeks passage without permits. As it was, the NOAA permit arrived one week, and the Australian permit within 24 hours of Australia Day, the Australian permit with the welcome message "Good Luck and calm seas." The latter were not to be [210].

A major concern was whether the distant stations would be able to detect our signal. Birdsall had made several estimates that varied from "undetectable" to "easily detectable." I cannot think of any other experiment we conducted where the outcome was so uncertain.

We arrived at Heard Island five days prior to transmission, with the three kilometer high glaciers atop Big Ben glowing over us (Big Ben has been climbed only once). The island was discovered by American Captain John Heard. For navigation we used the 1853 charts prepared by Mrs. Heard (they were excellent). The acoustic technicians aboard that *Cory* followed their usual practice of a five-minute equipment check just prior to zero hours. I had gone to bed in anticipation of a 24-hour day, when I was woken with an angry message from Metzger in Bermuda, "Received signal 12 hours prior to zero hours. What is going on?" I had hardly gone back to sleep when a second message arrived, this one from Birdsall at Whidbey Island near Seattle, Washington, having received our test transmission from the opposite way around the globe. So here was the answer to our concern, and it was not yet Australia Day. That moment was the high point of my career.

von Storch: I hope you slept well after that. And how did things go the following day?

Munk: They went well (Fig. 8.2). Our transmission was detected at all stations except by the Japanese station in the Tasman Sea, which was blocked by an uncharted seamount. After a week of transmission we were hit by a storm and all ten sources

Fig. 8.2 Walter arriving at Heard Island on Australia Day (26 January 1991)

were demolished. One was left lying on the sea floor. When I am asked, what is the longest you have ever had a source in the water, I say fifteen years and getting longer by the day. It was during the storm that I suddenly recalled that thirty years earlier we had drawn a crude weather map centered on some distant uninhabited island called Heard Island [85] (Fig. 8.3). The information came from tiny wiggles in a spectrum of waves recorded off San Clemente Island. And now we were here, where the swell begins.

The captain tried to cheer me up, "We had been told that the equipment could withstand any imaginable sea state. And we have gone to some length, to some discomfort, to test this statement." When I reported to Roger Revelle back at Scripps he responded, "I wish I were with you, and then again I'm glad I am not." After the storm abated, *RV Cory Chouest* beat her way westward to Capetown. One day out of Capetown, we made a landfall on Prince Edward Island calling on a meteorological contingent who had not had an outside contact for 14 months. (Among them were two professional sharp shooters with orders to eliminate the wild cat population that was eating the local birds.) Access was up a cliff via exposed rope ladders. On the way back a member of our scientific party refused to climb down; she had to be lowered by rope. I slipped returning to the boat and was saved from falling overboard by Elmer S. Hindman III of the *Corey* crew who grabbed the collar of my jacket.

Hasselmann: Did you learn something worthwhile?

Munk: I think so. The *Journal of the Acoustic Society of America* (*JASA*) devoted the entire October 1994 issue (95:4) to "The Heard Island Papers; a contribution to

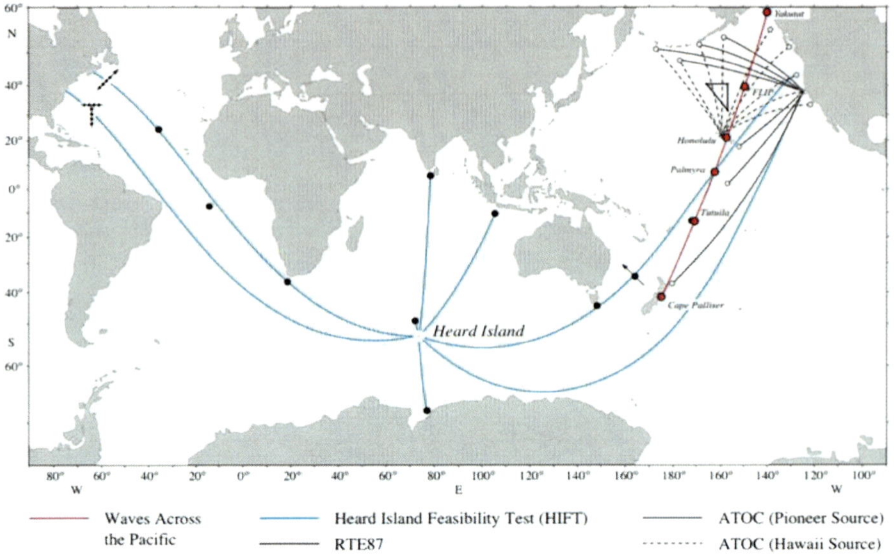

Fig. 8.3 A sequence of antipodal obsession. In 1964, waves from the southern hemisphere storms were followed along six stations (*red dots*) from New Zealand to Alaska. This established the origin of California's summer swell, some from as far away as the Indian Ocean. In 1991, a global array of hydrophones (*black dots*) recorded acoustic signals from Heard Island in the Indian Ocean, thus establishing the feasibility of global acoustic transmissions; after a week of successful transmissions the sources were demolished in a storm of the type that had been monitored from afar (and forgotten) twenty-seven years earlier. In a climate oriented basin-scale experiment led by Peter Worcester from 1996 to 2006, acoustic transmissions from Pioneer Seamount (*black lines, solid*) and Kauai, Hawaii (*dashed*) to an array of SOSUS stations monitored the decadal heat content of the North East Pacific using the methods of "Ocean Acoustic Tomography"

global acoustics." There were 17 contributions. Birdsall discussed signal-processing issues in the face of the enormous *a priori* uncertainties. During transmissions the *Cory* was underway at 4 knots into wind and waves to avoid being broached. It developed that the motion of the source ship (which we regarded as a necessary evil) was turned into a magnificent asset. Maximum Doppler is associated with transmission in the direction of the ship's course; any reduction is a measure of the geodesic launch angle relative to the ship's course. We found (i) the overall mean Doppler yielded accurate geodesic launch angles that agreed well with *a priori* calculations of the acoustic paths for all stations except those to the American west coast. For those the angles disagreed wildly. Evidently the direct path through the Tasman Sea is blocked, but a relatively weak signal reached the west coast passing east of New Zealand. (ii) Slight differences in Doppler between early and late ray-like arrivals at a fixed station are associated with the vertical inclination of the source angle and again the associated Doppler can be used for ray identification. (iii) Correlated phase fluctuations over the entire frequency band were traced to slight variations in the speed of the source ship (surge). Departures of order 5 m from a steady course (determined from the Ascension Island and Christmas Island records) are in accord with GPS (Global Positioning System) aboard *Cory* (taking advantage of the fact that GPS was on a war alert and its coarse acquisition (CA) code was NOT degraded in the usual manner.) So we can check the performance of a helmsman at 10,000 km distance (which is of absolutely no use). Nevertheless, we concluded that even if thc weather had permitted a stationary source, the information gained by (i) and (ii) would suggest using a moving source.

Kuperman reported on the stripping of acoustic modes, and the subsequent repopulation. The scattering into neighboring acoustic modes near sharp oceanic fronts (principally the Antarctic Circumpolar Front) and from bathymetric features intruding into the sound channel and the repopulation of previously dissipated modes are an important attribute of global acoustics which had not been anticipated. The subtitle to the *JASA* volume was, "A contribution to global acoustics." We might immodestly claim that the Heard Island Experiment initiated a new discipline.

The biological study was a late addition and severely limited by the short period available for the base study prior to the transmissions. Observations were restricted to times when the winds were weaker than force 8. The limited evidence suggests some behavioral changes in whales, but there is no evidence of distress.

von Storch: Would there be any point for a repeat?

Munk: Dushaw has suggested a repeat of the Perth–Bermuda transmission. The reduction in travel time due to global warming is estimated to be of the order of 10 s! A great part of the acoustic path is in the deep waters of the southern oceans where we have very little information. It would be lovely to have a single measure of global ocean warming over the last fifty years. But what measure? If the Keeling measurements of atmospheric CO_2 at a single point (Mauna Loa) can give meaningful information about a global increase by 6 gigatons of atmospheric CO_2 per year, why cannot measurements of ocean temperature at a single location (or along a single path) give a meaningful measure of increased joules of ocean heat content?

Hasselmann: Because trace gasses in the atmosphere are well mixed whereas heat in the oceans is not.

Munk: What is at the basis of this asymmetry?

Hasselmann: In the atmosphere, the residence time of greenhouse gases, in particular CO2, is large compared with the mixing time, whereas heat in the deep ocean is transported with little large-scale mixing on centennial to millennia time scales, so the temperature changes due to global warming are quite inhomogeneous. And superimposed on the global warming signal in the upper ocean is a significant decadal-scale natural variability.

Munk: We find that even within the relative confinement of the northeast Pacific, some of the acoustic transmissions show decadal cooling, in opposition to the overall trend. There have been many occasions where some striking ocean observations turned out to be only of local significance. With this in mind, when fifty years ago Keeling was about to commence his famous recording of CO_2 at Mauna Loa, I expressed my doubt whether this could ever give an indication of a global trend. Fortunately no one paid any attention, and the Keeling Curve is now the most famous time series in climate.

8.5 Whales

von Storch: You previously spoke about the problem of interference with marine mammals. When did this problem first come up?

Munk: The problem first came up during the planning stage for the Heard Island Experiment. In August 1990 we were informed that we required a permit from the National Marine Fisheries Service of NOAA. To make planning even more difficult, the Australian authorities who had previously supported the test plan without raising the issue of a permit, now decided that one would be required. The concern was that the level of acoustic sources was potentially a threat to marine mammals. This was unexpected; the question of a permit had never come up in 12 years of work in ocean acoustic tomography, albeit with less powerful sources. Moreover, oceanographers had long used much more powerful sources for seabed exploration.

We were required to charter a second vessel, the *RV Amy Chouest*, for *in situ* biological observations. Ann Bowles assembled in record time a team of three Australian and six American biologists. We worked out a protocol with the permitting authorities whereby, should there be any evidence of, or potential for harmful effects on marine mammals, the experiment would be delayed or cancelled. As it turned out there was no indication of harmful disturbances, and in some instances the animals even swam towards the *Cory* during transmission. The "biological add-on" certainly added to the expense and excitement of the feasibility test, yet it provided a welcome partnership with a devoted group of observers who worked under some very severe weather conditions.

von Storch: What happened after your return?

Munk: There was a lot of publicity. I have previously mentioned the article "Radau in der Tiefe" in the August 1991 issue of *Der Spiegel*. We now proposed a continuation of the ATOC (Acoustic Thermometry of Ocean Climate) experiments in the Northeast Pacific. At Heard Island we had transmitted 1 out of every 3 hours at a depth of 150 m using multiple sources with a combined 16,000 W (221 dB re 1 μPa) of acoustic power. The proposed ATOC schedule was six 20-minute transmissions every fourth day (2% duty cycle) at 260 W (195 dB) at 1 km depth. We were at sea installing a source when a front-page article in the *Los Angeles Times*[3]

[3] Paddock, R.: Undersea Noise Test Could Risk Making Whales Deaf. *Los Angeles Times* 22 March 1994, A1 (1994)

Fig. 8.4 Peter Worcester, Walter, and Carl Wunsch, authors of Ocean Acoustic Tomography in Venice, Italy at the OceanObs Conference (2009)

claimed that ATOC would lead to the death of 750,000 California Gray Whales. The problem was that the decibel scale is different in air than in water. 195 dB in air corresponds to 260 million watts (three million times the proposed 195 watts), which would indeed be fatal. I had a strained meeting with the reporter, Richard Paddock, and the error was later corrected (again on the front page).[4] But the outcry in the international press with literally many hundreds of articles (99% unfavorable) has led to years of frustration and legal proceedings.

Carl Wunsch, Peter Worcester, and I have summarized our findings in *Ocean Acoustic Tomography* (Fig. 8.4). Peter and I spent three months in Tasmania in 1989 making up our minds what was to be included (Fig. 8.5). This was my third (and last book) and again we chose Cambridge University Press as our publisher [225].

8.6 The Last Twenty Years

I have had an early experience with a project (Mohole) that failed because the principal investigators (including me) would not give it the time required to keep it afloat (perhaps we can discuss that later). I had walked away with the determina-

[4] Paddock, R.: Beneath the Not So Silent Sea. *Los Angeles Times* 3 May 1994, A1 (1994)

Fig. 8.5 Judith and Walter in Tasmania (1989). They spent three months working with the Worcesters outlining the Munk, Worcester, Wunsch book Ocean Acoustic Tomography

tion never again to participate in a project unless I was willing to give it the all the time required. And so Judith and I, and Pete Worcester, have participated in dozens of public hearings in Hawaii, California, and Washington. Let me tell you about a typical hearing.

v. Storch: Do we have to listen?

Munk: Please do. For example, a meeting was announced at 7 PM at some public high school gym on the Island of Kauai. There would be about 500 people, some of which had signed up to speak for 3 minutes. We had the first 20 minutes to present the subject. Then for about 5 hours there would be a hundred speakers, nearly all of them voicing opposition to the acoustic transmissions.[5] What I found the most difficult is not being able to respond when the opposition was based on a mistaken factual premise. Some time after midnight we were permitted to respond, but by then there were only a few hardy souls left in the hall. I remember Judith and I walking home through empty streets wondering what had been accomplished.

But we did keep Acoustic Thermometry afloat. Under Pete's leadership the technology has dramatically improved. Pete deployed a source on Pioneer Seamount off the coast of California, and later a source north of Kauai, Hawaii. Transmissions were to a dozen SOSUS receiver arrays in the Northeast Pacific. Permits for the

[5] We will not forget the appearance of Darlene Ketten at many of these meeting. Darlene is on the Woods Hole faculty and perhaps the leading student of the hearing of marine mammals. She never hesitated to present pertinent experimental results to an unfriendly audience.

sources included a sunset clause. We ended up with a decade of northeast Pacific transmissions.

The conclusions are that (i) acoustic thermometry does a very good job of estimating the heat content of the northeast Pacific, (ii) ten years was not long enough to provide a good measure of long-term warming in the presence of intense decadal variability, and (iii) comparison of the measured travel times of individual transmissions with those derived from modeling, Argo Float measurements plus satellite altimetry generally shows good agreement, but occasional sharp disagreements (which are not understood).

Klaus has heard some of this before; when I was given a chance to vent our frustrations in the Reimar Lüst Lecture [238d] "Listening to Ocean Climate" at the 25th anniversary of the Max-Planck-Institut fuer Meteorologie.

Hasselmann: Yes, I have heard this before. And I am impressed by your persistence in the face of much adversity – and the advances you have nevertheless achieved! And how would you assess the future role of Acoustic Technology in global climate studies?

Munk: Grim! I will ask you to listen to my prejudiced summary assessment. We all agree that climate change is a leading problem of today. About two-thirds of the incremental heat content is stored in the oceans, and needs to be measured for any sensible bookkeeping. Further, substantial global temperature variations with decadal time constants are superimposed on greenhouse warming, and these need to be understood, predicted and subtracted from the overall warming in order to convince society to do something about the anthropogenic component. I suspect that Enso-like ocean variability is at the root of some of this and will require ocean temperature monitoring on a basin scale.

Hasselmann: I agree with you entirely. But what is to be done?

Munk: Oceanographers are monitoring the ocean now with a combination of Argo floats, satellite altimetry and modeling. I believe we would do significantly better if OAT[6] were a component. OAT has clearly lost out in the competition with Argo. Argo has the advantage of small start-up costs (though the overall cost may be equivalent). OAT has the advantage of an inherent averaging, vertically through the total ocean column, horizontally on a basin scale. Twenty-five pelagic moorings with appropriate global distribution would do the job. NSF at one time had set aside funds for a global mooring system (as successor to weather ships), with provisions for power, data transmission and time keeping (this would take care of most of the cost of OAT), but has since withdrawn the mooring system into coastal waters.

This leaves the marine mammal problem. I believe that OAT can do a superior job, and with proper precautions will do no harm to marine life. But given the opposition

[6] For Ocean Acoustic Tomography. As a member of the AAS (Anti-Acronym Society) I have tried until now to avoid this ugly step.

of environmental groups, NSF is (and even some of our supporters are) afraid to touch it.

von Storch: And what about the future?

Munk: Only God knows. Let me remind you of the Mohole project. There, after having solved some very challenging technical problems, the project was terminated on account of the incompetence of the principal contractor. But ten years later it arose spontaneously in the guise of the immensely successful Deep Sea Drilling Project (DSDP). OAT is still waiting.

Chapter 9

Mohole 1957–1964

Hasselmann: Oh yes, the Mohole business, I well remember the drama of its later stages while I was at IGPP in the early sixties.

von Storch: I don't know the name.

Munk: One of the major boundaries in the solid Earth is the transition from crust to mantle. This is associated with a change in seismic properties at 30 km under continents and 5–10 km under the oceans floor. It was discovered by Professor Andrija Mohorovičić, a Yugoslavian seismologist, whose name no one could pronounce.

In spring of 1957 the initial NSF panel on Earth Sciences had just sat through two days of reviewing sixty-five proposals. We were recovering over a drink when the question came up of what would be the project that could lead to the greatest advance in understanding the Earth, cost being no object? Harry Hess and I suggested drilling the "Mohole" through the sea floor to collect a sample of the Earth's mantle. The required development of keeping a drilling vessel in place and, if necessary, of re-entering the hole, seemed within reach of existing acoustic technology.

We put together an informal group to plan a feasibility test: Revelle, Hess, Joshua Tracey, Maurice Ewing, Harry Ladd, Munk. We chose an unlikely group for administrative support: AMSOC.

von Storch: Who is AMSOC?

Munk: The American Miscellaneous Society "for maintaining cooperation with visitors from outer space and for informing animals of their proper taxonomic positions" was founded by Gordon Lill, John Knauss, and Maxwell, all from the early ONR. The Society is internationally known for the Albatross Award, a stuffed bird presented at irregular intervals to an oceanographer for some "unusual achievement" (I am a proud Albatross Laureate). There are no officers, no dues, no bylaws. AMSOC submitted a formal proposal to NSF. The Foundation did not decline outright, but suggested that it would entertain a proposal from the same group if they attached themselves to a more substantive organization. And so there came about the

H. von Storch, K. Hasselmann, *Seventy Years of Exploration in Oceanography*
DOI 10.1007/978-3-642-12087-9, © Springer 2010

AMSOC committee of the National Academy of Sciences (NAS). I.I. Rabi, who reviewed the proposal on behalf of the Academy Council, remarked: "Thank God, we're finally talking about something besides space."

By 1961 we were aboard the CUSS I (named for the Continental, Union, Superior, and Shell Oil Companies) drilling off Guadalupe Island in 12,000 feet of water (Fig. 9.1). Willard Bascom was in charge working with Ed Horton and Francois Lampietti. The drilling vessel was continually "underway," driven by four large outboard propellers to maintain a fixed position relative to three sonic bottom transponders, the first demonstration of "dynamic positioning."[1] In spite of foul weather the drill penetrated 560 feet of sediment and then a few feet of basalt. John Steinbeck and Fritz Goro were along to record the event for *Life* magazine. The test was completed on time and within the allotted budget of $1.7 million.

When Bascom wired NAS President Detlev Bronk that we had reached basalt, we took it for granted that Mohole was in the bag. Little did we realize that from this moment on the project was doomed.

von Storch: Why is that?

Munk: What happened was altogether unsuspected. Industry, which had been on the sidelines until the Guadalupe drilling, now showed a sudden interest. When NSF held a briefing in July 1961 preparatory to appointing a Prime Contractor, 200 people representing 84 companies attended. By the time of the deadline – September – ten proposals had been received. The proposal by Socony Mobil in partnership with General Motors, Texas Instruments and Standard Oil of California was rated first by the NSF Evaluation Panel ("... in a class by itself"). A proposal by Brown & Root, a Houston based construction firm with no offshore experience was rated fifth. When questioned about the lack of experience, Herman Brown is said to have replied, "I can always hire an acre of engineers."

The rest is history.[2] NSF Director Waterman chose Brown & Root. Three years later, with $50 million spent and no results, the project folded; Mohole had become No Hole. We have never found out how Brown & Root had been chosen, but the rumor mill attributed it to the fact that Herman Brown was a friend of President Johnson. Further, we had all taken for granted that the Bascom group would be involved, but after an initial period they were removed.

von Storch: And what is the lesson here?

Munk: We must all share the responsibility for the Mohole debacle. It was foolish to expect that we could go home and do our favorite science and expect the project to proceed according to our plans. I walked away with the determination never again

[1] This was a significant step towards developing the technology of offshore oil production.

[2] Solow, H.: How NSF got lost in Mohole. *Fortune* May, 138–141, 198–199, 203–204, 208–209 (1963)

Fig. 9.1 Aboard CUSS I east of Guadalupe Island during Project Mohole (1961). This was the first attempt to drill into the deep sea bottom and required the development of "dynamic positioning" for the drilling vessel using acoustic bottom transponders. (*Left* to *right*: Roger Revelle, Willard Bascom, Walter Munk and Gustav Arrhenius)

to participate in a project unless willing to give it the time required. The Mohole experience accounts how we reacted to the opposition to Ocean Acoustic Tomography. Over tens of years there have been many dozens of meetings, in Washington, in California, in Hawaii, where we responded for many hours to charges that OAT was endangering marine life. Some charges were reasonable, and we have modified the experiments accordingly; most of the charges made no sense. Pete Worcester has carried the brunt of this.

Hasselmann: So that is what you had in mind when you spoke about the whale problems encountered in Ocean Acoustic Tomography. But the two situations are really very different. OAT is now a functioning technology, even if it did not achieve the prominent role you and Carl Wunsch had envisioned. And with regard to Mohole you started out by having fun playing games with the American Miscellaneous Society; whoever heard of a government agency (a funding agency in particular) with a sense of humor? How could you expect to have NSF take you seriously later on?

Munk: ...

Hasselmann: And then what happened?

Munk: We have already spoken of the death and resurrection of the Mohole project: the hugely successful Ocean Drilling Project. Will there be an equivalent history to Ocean Acoustic Tomography? We are waiting.

Chapter 10

The Wobbling Earth 1950–1960

Hasselmann: Your early work was on water waves, and your recent work on acoustic waves, both dealing with the fluid ocean. The Mohole discussion dealt with solid Earth geophysics. Tell us about other non-fluid topics you have worked on.

Munk: I spent a decade on problems associated with the irregular rotation of the Earth. It came about by accident. I had learned about the high-altitude atmospheric jet stream from the Finish meteorologist Eric Palmen (whom I had met through C.-G. Rossby).

von Storch: I believe it was discovered by the U.S. Air Force in WWII when they noticed it took longer than expected to fly westward to Japan.

Munk: It parallels Ben Franklin's discovery of the Gulf Stream when he noticed that mail from England to the United States took a day longer than on the reverse route. Palmen found that the high altitude jet contained most of the atmospheric angular momentum. And Victor Starr in 1948 had noted a seasonal variation (only partially cancelled by the summer hemisphere jet) must be accompanied by "undetectable inequalities in the rate of the Earth rotation...," the net angular momentum of the Earth being conserved. How big is undetectable? A stronger winter jet would be associated with a longer winter length of day (lod). I started reading about clocks and found that M. Nicolas Stoyko and Mme A. Stoyko of the Bureau de L'Heure in Paris had discovered in 1936 that the lod in January exceeded the lod in July by 2 ms. This was based on precision pendulum clocks. A very simple calculation showed that this measured variation agreed in amplitude and phase with that inferred from the seasonal wind variation [21].

Jeffreys had noted as early as 1916 that a seasonal shift in air mass associated with the high pressure in winter over Siberia must lead to wobble of the Earth relative to a rotation axis fixed in space. Wobble and spin are the x, y, and z-components of a single vector of Earth rotation. It is not *a priori* obvious that wobble should be controlled by mass shifts and spin by the velocity variations. When I mentioned

H. von Storch, K. Hasselmann, *Seventy Years of Exploration in Oceanography* 71
DOI 10.1007/978-3-642-12087-9, © Springer 2010

the lod result to Jeffreys during my 1955 sabbatical in Cambridge, England, he responded with a loud, "Oh No!" It was his only expression of surprise in many years of association.[1]

There was a curious dichotomy between the x, y-community (International Latitude Service) and the z-community (The Time Standard people) even though they were measuring the effects of the same geophysical processes. It would seem worthwhile to study the processes side by side.

Hasselmann: So that is what prompted you to dive into this subject?

Munk: Yes indeed. From the annual variation I went into the Chandler Wobble, the 14-month free rotation of the Earth. This is the source of the one cm pole tide[2] (the only tidal component NOT caused by the gravitational attraction of Moon and Sun). I was giving a talk on the pole tide at Harvard when I was severely questioned by an undergraduate in the first row. A year later I gave a similar talk at Cambridge, and there he was again in the front row, taking me to task for not having responded to his previous criticism. When you can't fight them, join them. And that is how Gordon MacDonald (the undergraduate critic) and I got started on writing *Rotation of the Earth; A Geophysical Discussion* [72].

We dedicated our book to Sir Harold Jeffreys. Jeffreys worked at the Met Office during the WWI and his interests shifted to Geophysics. His book *The Earth* had gone through three editions (1924, 1929, 1952). At the time we published *Rotation of the Earth*, *The Earth* had become the bible of theoretical Geophysics. Unfortunately he refused to recognize revolutionary developments in what is now known as Plate Tectonics, and the three subsequent Editions published in 1962, 1970, 1976 (reprinted 2008) increasingly lost their relevance. I wish Jeffreys had stopped with his third edition.

Hasselmann: Neither you nor Gordon really knew anything about Astronomy. There must have been some resentment by the traditional astronomic community.

Munk: There is above all the problem of learning a new language. In the United States time keeping is done at the Naval Observatory in Washington D.C. William Markowitz was our nation's timekeeper, and he took a dim view of the oceanographic intrusion into his domain.[3] There is some resemblance between the communities of position astronomers and tidal analysts. We found that some topics can

[1] In 1914 when my mother was a student in Cambridge with a major in Botany, Jeffreys was Reader in Botany, His war work in the Meteorological Office turned his interest to geophysics, and for many years he dominated the field, with some emphasis on problems in Earth rotation. He is known for having calculated Earth wobble due to leaves falling to the ground in autumn, and related seasonal processes (this intriguing calculation is probably the only geophysical application Sir Harold has been able to make of his early days as a botanist).

[2] Carl Wunsch wrote his PhD thesis on long period tides. Carl was a student of Henry Stommel who detailed Carl to La Jolla for information on the odd topic of the pole tide.

[3] My 65th birthday celebration was published (SIO Reference Series 84-5) with the title line, "It's the water that makes you drunk." W. Markowitz told of the logical scientist who set out to

be clarified by including consideration of the underlying noise continuum. Gordon and I then followed the trail from the "high-frequency" annual terms to irregularities of ever longer periods.

Hasselmann: So here again you followed a pattern of going from high to low frequencies. This is the pattern you followed in your wave work, and much later in acoustic tomography. I wonder whether there is an advantage of exploring in the direction of increasing frequency rather than the other way around?

Munk: I had not realized until just now that this is a repeated pattern in my work. Whether there is an inherent advantage, I don't know. I suppose it is a way of avoiding being fooled by high frequencies appearing under the alias[4] of low frequencies.

Hasselmann: What were some of the lower frequency phenomena?

Munk: There is a large decadal variation in the lod, which goes under the name "The Great Empirical Term of Newcomb." Relative to some mean, the lod varied from

$$1970: -4\,\text{ms}; \qquad 2005: +3\,\text{ms}; \qquad 1930: -2\,\text{ms}.$$

This is larger than the annual amplitude of order 1 ms. The cumulative time error is of order 10s! This term is presumably associated with a variable rotation of the fluid core of the Earth, and there is some evidence provided by comparison with the westward drift of the Earth's magnetic field (I have not kept up with recent literature). Newcomb was Astronomer Royal at the turn of the century. We had the pleasure of discussing this problem with his successor Sir Harold Spencer Jones at Herstmonceux Castle.

On a century time scale the variable rotation is dominated by the effect of tidal friction. We are so fortunate to live in an epoch where the angular diameters of Sun and Moon are nearly equal, so that the occurrence of an eclipse is a remarkably accurate statement of the positions of Sun and Moon. Halley in 1695 compared his "modern" observations with positions derived from ancient eclipses and concluded that the Moon was being accelerated. The associated numbers are astounding: ancient eclipses occurred hours later, or thousands of kilometers eastward, of their expected times and positions. Since the study of ancient eclipses requires competency in both astronomy and antiquities, the field has never been overcrowded.

Here I have referred to the z-component of the vector of Earth rotation. There is interesting phenomenology associated with the x, y-components. For example, the

---

determine the cause of drunkenness. On successive days he consumed large qualities of scotch and water, bourbon and water, and vodka and water. Becoming drunk in each case he concluded that the water was responsible. Markowitz went on to find a parallel in the authorship of a number of then recent papers on the rotation of the Earth [49, 60, 72] singling me out as the common ingredient in the infiltration of geophysicist into what had been the domain of astronomers.

[4] This is one of many examples where Tukey introduced some highly descriptive terms into the language of spectral analysis.

pole of rotation is now moving towards Greenland at the rate of 25 feet per century. The problems of Polar wandering on a geologic time scale has led to heated discussions by scientists and crackpots alike. The discussion is now dominated by the paleo-magnetic evidence that shows the large relative movements of continents. I am impressed by the changes in fashion of what is believed to be true at any one time. Physicists gave decisive reasons why polar wandering could not be true when it was weakly supported by paleoclimatic evidence; now that rather convincing paleomagnetic evidence has been discovered they find equally decisive reasons why it could not be otherwise.

von Storch: I can understand why you have found this subject so fascinating. But has it led to a solid advance in what we understand about the Earth?

Munk: I think so, and here is why. The diversity of the subject is appalling, touching wind and air masses, atmospheric, oceanic and bodily tides, sea level and motion in the fluid core. But the information is limited to certain integral quantities taken over the entire globe. This is the weakness of the method – and its strength. In principle such integrated quantities can be evaluated by appropriate summations of individual station values. For competitive accuracy one usually finds that the stations are too unevenly distributed, and too few. This is true now; we doubt whether it will ever be any different.

Hasselmann: A great way to learn about the subject is to look at it from diverse points of view. If nothing else, your sojourn from oceanography must have been a fine learning experience.

Munk: It was. But it was something else also. The first draft was written during a year in Cambridge. It provided the opportunity of getting to know some extraordinary people, Teddy Bullard, G.I. Taylor, Harold Jeffreys, and others. And I fell for a Cambridge-like existence, heavily based on individual scholarship. By the time I returned to Scripps I was uncomfortable with our emphasis on coordinated big projects.[5] I thought seriously about an offer at chairing Geophysics (or whatever it was called) at Harvard. Judith said, "Why don't you stop bitching and start a new group *here*?" She added, "And I will build you the laboratory."

---

[5] As it turned out thirty years later, I was to be the organizer of the then largest Scripps project, the Heard Island Feasibility Test.

Chapter 11

Institute of Geophysics and Planetary Physics (IGPP) 1962–Present

Hasselmann: And exactly how did things get started with the IGPP?

It was Roger Revelle, as always. There had been an offer from Bob Shrock, chairman of Geology at MIT, which I was about to accept. There was another offer from Harvard. Roger listened to my objections (which he did not share) about large projects and then said, "Is there anything you could do in Cambridge, Massachusetts that you could not do in La Jolla?"

Hasselmann: So you tried to create a bit of Cambridge in La Jolla?

11.1 The Cambridge Connection

Munk: You might put it that way. By the time IGP (Institute of Physics) was founded (it became IGPP when Gordon MacDonald added Planetary Physics) Judith and I had spent two Guggenheim Fellowships in Cambridge, England, 1955 and 1962. There was to be a third Cambridge Sabbatical at Churchill College in 1981, and in 1986 I received a Cambridge Honorary Degree.[1] Cambridge has played a major role in our lives. And my mother spent the happiest two years of her life playing

[1] Prince Philip, Chancellor of Cambridge University, officiated at the ceremony. The careers of the recipients were traditionally reviewed at some length in Latin, often evoking loud laughter. I recall one instance when Prince Philip and I glanced at each other, and then joined in the laughter. Evidently we were the only ones present who did not understand Latin. The night before we had been at a small dinner party at a Master's Lodge where Judith sat next to Prince Philip and talked about sailing races for the handicapped. We had walked over from the Garden House Hotel and I was surprised there was no sign of any security presence when we walked up to the Lodge to ring the bell. Prince Philip opened the door with the words, "I bet you want a drink." We had met in 1983 when the Queen and Philip visited the Scripps Institution. Director William Nierenberg attended the Queen, and I was assigned the pleasant task of showing the Institution to Prince Philip. Nierenberg had decided that he should be briefed on the subject of climate (this was many years before climate had become a subject of public interest). I took Prince Philip to the office of

DOI 10.1007/978-3-642-12087-9,

hockey at Newnham College from 1912–1914, but had to return to Austria when WWI started. At that time it was most unusual for girls from the Continent to study in the UK. Her major was in botany, and her tutor was Harold Jeffreys, then a Reader in Botany.[2]

Hasselmann: You have told me that you have three heroes: Sverdrup, Revelle, and G.I. Taylor. Did you meet Taylor at that time?

Munk: Yes. The three have been my role models: my teacher Sverdrup, my mentor Revelle, but if you have to use the word "genius," Taylor is the one. I can imagine writing some of the papers Harald has written, or fighting some of Roger's battles, but G.I. ..., never. Whether it had to do with wind shear, turbulence, a mushroom anchor, crystal dislocation, a combination of brilliant insight plus careful laboratory measurements would yield fundamental truths. In 1986 I participated in a celebration of his hundredth birthday (he was then ten years gone) and said that everything G.I. had touched turned to gold. Not so, said George Batchelor, who took me to an attic which stored folders upon folders of work that G.I. had abandoned. Evidently if an idea did not yield significant results, theoretically or experimentally, within a month, G.I. would set it aside. Letting things go is not a talent I share.[3]

von Storch: I have heard you speak of Sir Edward Bullard as one of the people who made IGPP what it is.

Munk: Yes, Teddy played a major role. In his later life he spent winters at IGPP and he died in La Jolla in 1980. By the time I met him, he had truly transformed the field of geophysics. In the early 1940's all the major geophysical tools, based on seismology, gravity, magnetism and geothermal heat flow, had been developed for use on land, and their application for use at sea was considered to be impossibly difficult. Bullard played a significant role in the adoption of all four methods for work at sea; in the cases of heat flow and magnetism he played the leading role.

Professor Jerome Namias, a pioneer in climate science, and was waiting outside the door for what I expected to be a short discussion. But of course Namias could not stop talking about his favorite subject. An hour later Prince Philip came out wiping his brow, "I'm glad we don't have to worry about climate in Britain, we already know it's going to be beastly."

In 1997 during the last trip of the *Royal Yacht Britannia*, an official reception was held in Hong Kong. My brother Alfred, who had a distinguished career as an economist with Standard Oil of Indiana and rather looked down on his brother's impoverished academic career, was visiting there and attended the reception. When his name was announced, Prince Philip asked, "Are you related to Professor Munk?" Things have never been the same.

[2] Mother was with us in 1955 when Sir Harold came for tea. He took one long look at mother and said "Brunner" (her maiden name).

[3] G.I. once took me to dinner at Trinity College, and spoke of some recent work by Michael Longuet-Higgins on the role of collimation in turning a confused, offshore sea into coherent long-crested breakers. As wave crests turn parallel to shore by wave refraction in shallow water, the subtended angular beam is narrowed and the waves become more "long-crested." "I don't think that's the whole story," G.I. said. He had in mind the non-linear capture of short waves by longer, faster waves coming from behind. The problem is still under consideration.

Fig. 11.1 Teddy Bullard, a Cambridge University Demonstrator, taking gravity measurements in South Africa (circa 1931)

Today all these methods are routinely used at sea, and in fact their most successful application has been at sea, not on land. Part of the reason for this reversal is, I think, because the deep sea is such a benign environment, so well thermostated. But more important, because the secrets of plate tectonics were hidden under the deep sea and had to be unraveled by the geophysical measurements at sea. Land geologists could have banged their hammers on rocks for 10,000 years without having a clue about the dynamics of continents and ocean basins.

Nearly all the members of the Bullard Laboratory on Madingley Rise have spent time in La Jolla; and nearly all of us at IGPP have spent some time at the Bullard Laboratory. Robert Parker, one of our most distinguished faculty members, is a Bullard product.

Hasselmann: What was your personal interaction with Bullard?

Munk: It was so much fun interacting with him. At one time I had just written the introduction to a volume dedicated to Bullard when he came up for another honor, and I was again asked to summarize his career. I told Teddy I was tired of writing about him; would he not enjoy saying what he thought was really important in his career, and I would submit it under my name. He wrote an essay about the things he remembered, such as coming back from the War and having to wipe the filthy floor in his Cambridge Laboratory. The manuscript was returned some months later with most of it crossed out in thick red ink. I protested to the Editor who replied, "We could not possibly publish what you had sent, Sir Edward would be gravely offended." I prevailed, and the essay was published. Teddy got such a kick out of this that he eventually spilled the beans.

When we were about to move into the IGPP Laboratory, the University architects attached four-digit numbers to each of the offices. But who wants to sit in room 4257? For identification, we attached photographs of leading geophysicists on the doors. One of them is of Teddy in the nude (Fig. 11.1), doing magnetic measurements in Africa. I asked him if he would mind the use of his picture, he said "No, as long as I do not have to sit behind it." In reply to your question, Hans, IGPP indeed carries Teddy's imprint, but not so much because of his very considerable intellectual achievements, but because he was so much fun.[4]

Hasselmann: What about your students?

Munk: They are an indispensable part of IGPP history. Let me mention two students who became life-long friends. Jim Cairns came to us from the Point Loma Navy Laboratory and received his degree in Physical Oceanography in 1974. From the very start he was pre-occupied with electric connectors in an ocean environment. Existing connections were made in an air environment protected by a leak-proof housing built to withstand large exterior pressures. Jim asked, "Why not make the connections in non-conducting fluid at ambient pressure?" and spent his life designing wet-mateable connectors. In 2009 when he sold Ocean Design Inc. (ODI) to Teledyne, it held virtually a global monopoly in underwater electric connectors. Judith and I had invested $10,000 in Jim in the early 1980's when he was dirt-poor. Last year upon the sale of ODI I retrieved a million dollars. We have stayed in close contact. I recently had a splendid visit with Jim in a 12th century nunnery near Urbano, Italy, which Jim and his wife restored for their residence.

Another student, Giuseppe Notarbartolo di Sciara, had been introduced to us by my Italian cousins. In 1974 when he first came to Scripps to study marine biology, he stayed with us for a few days while he was looking for lodging. As it turned out, he was to stay with us for seven years, until he received his degree in 1980. The Notarbartolo's played a distinguished role through centuries of Sicilian history. When Giuseppe arrived he was an advanced diver and held a captain's license. While working alone from a small boat off the coast of Baja California collecting data on manta rays, the "beautiful flying saucers in the Sea of Cortez" as he called them, his boat caught on fire and sank. Giuseppe barely made it to shore in a dingy, saving only his passport and a draft of his dissertation (the dingy is featured in the Seiche garden). He discovered a new species of manta rays (Fig. 11.2), which he named in

[4] This gives me the opportunity of recalling something that did not happen (Fig. 11.3). Teddy and I were driving through the English countryside when we passed a farm with a most remarkable contraption; a large piece of machinery with a collection of wheels and cylinders on all sides. We could not think of any possible function. Teddy took a careful photograph. At the time Members of the National Academy of Sciences could publish in the Academy Proceedings without review. Our intent was to submit a joint article with the photograph as Fig. 11.3 and legend "see text," but no further reference to the figure.

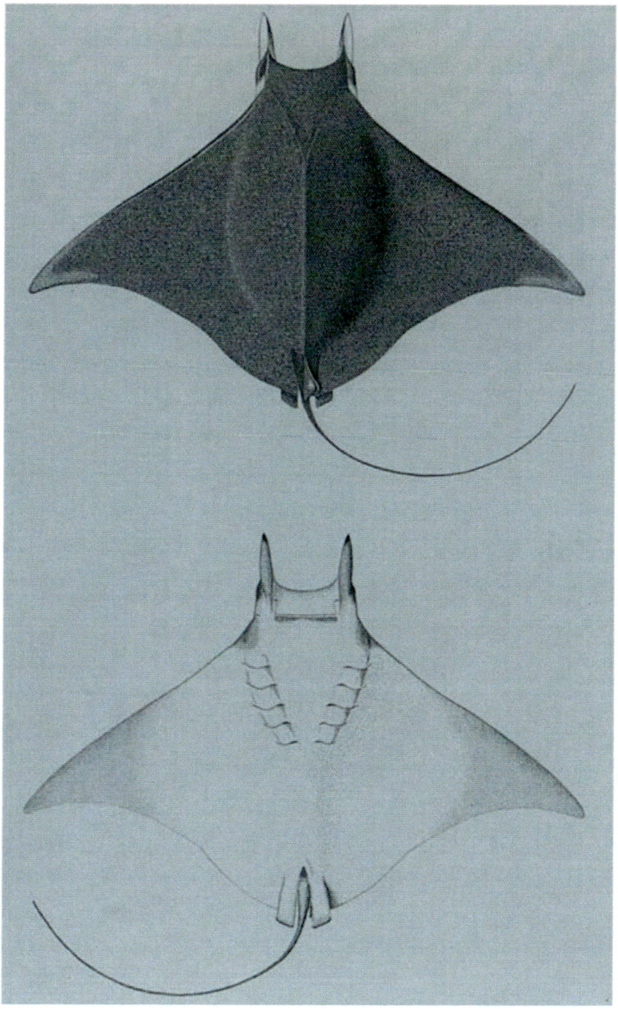

Fig. 11.2 *Mobula munkiana*, a species of manta rays discovered by Giuseppe Notarbartolo di Sciara in the Gulf of California in 1979

my honor, *Mobula munkiana*, and best of all asked me to best man at his wedding. Last year I spent a week at his home on the Greek island of Patmos.

A highlight of my career has been the time spent with my students as advisor on their Ph.D. committee, as participant in their adventures in research, and ultimately as their colleague and friend. Starting in 1949: William Van Dorn, Charles (Chip) Cox, Gordon Groves, Earl Gossard, June Pattullo, Arthur Maxwell, John Knauss, Gaylord Miller, Brent Gallagher, Mark Wimbush, James Irish, James (Jim) Cairns, Gordon Williams, Peter Worcester, John Spiesberger, Mike Brown, Donald Altman,

Fig. 11.3 "Look out below." See text

Jeffrey Nystuen, Bruce Howe, Brian Dushaw, and Werner Morawitz. I have learned as much from them as they have learned from me.

11.2 Finding the Faculty

von Storch: Lets get back to IGPP. How did you get started?

Munk: We started as a component of a state-wide IGP centered at UCLA, formed with the encouragement of Sverdrup and Rossby. There was an immediate problem as to whom I was to report to, UCLA director Louis Slichter, or Scripps director Roger Revelle. I remember a tense discussion between the two Directors. My loyalty clearly went to Scripps, but the problem never quite went away. At one time, Nobel Chemist Willard Libby who had succeeded Slichter as director, came to La Jolla to order me to report to him, and to take steps towards taking over some of Scripps, since Oceanography was a subset of geophysics, and not the other way around.

von Storch: What was your intended mission?

Munk: We never came even near to writing a "mission statement." But it as a matter of shear luck that forming our geophysical laboratory within an oceanographic institute pretty well overlapped with the plate-tectonic revolution. That gave us a general direction. I spent my time on appointments and promotion and on very little else; very little on what is usually called *Administration* and certainly not on money matters.[5] This permitted me to go on with research and teaching for the 23 years I served as Director of the La Jolla Laboratories of IGPP (1959–1982). I believe that IGPP faculty and researchers regarded me as a colleague rather than as Director, a crucial distinction. And so it has been with successive Directors.

Hasselmann: Well, that I can only confirm. I came to IGPP in 1961, just after it was created, and found the atmosphere extremely stimulating and exciting. In fact, I found your description of how Cambridge inspired you to create an institute founded on individual scholarship quite interesting, as I had exactly the opposite impression when I visited Cambridge after having spent some years at IGPP. I found Cambridge pleasant, but rather uninspiring in its tradition-bound, rather reserved individualism, in contrast to the excitement of the many seminars, discussions and spaghetti-and-wine parties in your home in those first years of IGPP. And it was essential, of course, for the spirit of IGPP that the director was having as much fun as everybody and was participating in a team effort.

When I became director of the Max Planck Institute of Meteorology in Hamburg in 1975, after experiencing other institutions as Doherty Professor at the Woods Hole Oceanographic and director of the Geophysics Institute of Hamburg University, it was always IGPP that I turned to as my role model of how to run an institute. Fortunately, the Max Planck Society has the same philosophy as yours, the directors should be relieved of administrative duties (with the support of the central Max Planck administration in Munich) and should be free to do good research. But how exactly did you go about making the appointments?

Munk: In a very unstructured way. Let me give you one example. I met Klaus Hasselmann at the Ocean Wave Conference in Easton in 1961, where Klaus presented his solution to the nonlinear interactions between wave components. Klaus introduced his talk with the following statement, "Basically, I solved this problem to relieve my frustrations at not being able to solve the turbulence problem." The room was filled with people who had for years tried to solve the wave interaction problem, and they were not exactly ecstatic with this statement of a twenty-nine year old. Life was simple in the early sixties, and I was able to offer Klaus an Assistant Professorship before the conference had closed.

Hasselmann: That is not exactly what I had in mind when I asked of how you went about making appointments.

Munk: Roger allotted us 4.92 FTE's (Full Time Faculty Equivalents), and by sharing some of these with appointments on the newly formed upper campus of UCSD we

[5] It was extremely fortunate that we came to maturity during the heydays of a basic science oriented Office of Naval Research. One day I received $50,000 in excess of a proposal to ONR for my personal research. I called my project officer, who said, "Yes, I know, we thought you could use some extra support for starting the new Institute."

were well on the way. (I have no idea where the 4.92 number came from.) Among the early appointments were George Backus who came from MIT, Freeman Gilbert from Texas Instruments, and John Miles from UCLA (via Australia). All of them eventually became members of the National Academy of Sciences. Some other appointments were not so good.

We never made our appointments in a structured way. We had no formal committees to choose candidates. We were looking for that person with that extra spark. I was very much under the Cambridge influence: Bullard, Jeffreys, Taylor. And I remembered a story, possibly apocryphal, about C.-G. Rossby, [6] the leading meteorologist of his time. Anyway our process of appointment was unstructured, and we were very, very lucky.

Hasselmann: Well, maybe it was luck, and maybe it was good instincts and judgment. This is, of course, the risk you take when appointments are made without the encumbrances of committees. We have the same uncomplicated procedures at Max Planck.

Munk: Years later Judith and I were reading about the resignation of the then President of the University of California. He was asked by a reporter, "What made you think of resigning at this time?" The reply, "I thought I had better think of it before others think of it." We took a long look at each other. Next week I invited all of IGPP to lunch and announced that Freeman Gilbert would take over. I had arranged to go to sea that very afternoon, and by the time I returned everything was going smoothly. Freeman was succeeded by John Orcutt (former Annapolis graduate), then Bob Parker (Bullard product), Cathy Constable, and today Guy Masters. If I may brag a little, I count eight of us at IGPP who are (or were) members of the National Academy or the Royal Society, or both. Seeing IGPP develop has been the most rewarding experience of my career.

11.3 Building the Laboratory

Hasselmann: Yesterday we walked up from Scripps Pier on Biological Grade past Scripps Library to come to a sign:

Munk Laboratory ← → *Revelle Laboratory*

[6] Rossby returned to his native Sweden to found a famous Meteorological institute. One of his appointments was P. Welander who had written some of the basic papers on the thermocline formation. As I remember the story, he encountered Welander in the hall, "And how is your father?" Welander replied, "There must be some misunderstanding, my father died many years ago." It turned out that there were two Welander's, and Rossby had appointed the wrong one. It was also his best appointment.

and there were the two red-wood buildings, distinct and yet harmonious. You must have been very pleased to have your association with Roger memorialized in this fashion.

Munk: Judith and I were very pleased. The association with the Revelle family is a central theme in our lives. Roger once said he did not care much for having a building named for him, but in fact he was quite pleased.

von Storch: Was it difficult to raise the money?

Munk: Not really. We had a number of Brownie points to start with. The University earmarked $486,000, half the estimated cost, provided we could raise the other half. Next we received $243,000 from the Air Force Office of Scientific Research (AFOSR), provided we could come up with the remaining half. Next $121,500 from the newly formed National Science Foundation, and then $60,750 from the Fleischmann Foundation of Nevada. You will note the resemblance to a geometric series $\frac{1}{2}+\frac{1}{4}+\ldots$ with a sum of 1 but only after an infinite number of terms. A timid suitor was once asked how he had approached his girlfriend. "Well," he said, "she let me come half way, and then half the remaining way...." "But then you will never get there," said the questioner. "But I got close enough for all practical purposes." We got close enough when we were looking for a remaining $15,000. An officer from the Research Corporation had come all the way to San Diego to decline our proposal because he couldn't see how $15,000 could possibly make a difference. We did not complain and took him to lunch. By the time his train reached Los Angeles he had changed his mind.

Hasselmann: I am surprised that the Air Force would lend support. I thought your connections were entirely with the Navy.

Munk: That goes back to the very start of our discussion when we spoke about oceanographers and seismologists being unfamiliar with power spectra. I hope my memory serves me right in the following account. The United States was negotiating a nuclear test ban with the Soviet Union, and there was a need for an appropriate monitoring system. Edward Teller was afraid of the Russians decoupling the explosions in underground cavities. America had proposed the "Geneva Network," consisting of pen and ink (!) recording seismometers of appropriate sensitivity. The original requirement was in terms of millimeter of pen displacement per kiloton of explosive energy at some specified range. At one point President Eisenhower had to withdraw from the agreement he had already signed because (if I remember the words) "... he had been misinformed by his scientists." I served on a second committee under AFOSR sponsorship which redefined the requirements in terms of a signal-to-noise ratio. One Sunday morning the Director of AFOSR, with two young Colonels on his sides (looking more like college students than senior officers) came down the steps to our house, "I understand you would like to start a new Institute...."[7]

[7] I had met AFOSR Director some years earlier when I was looking for an opportunity to photograph sun glitter as a means of estimating the probability distribution of sea surface slopes. If

Fig. 11.4 The IGPP Laboratory (designed by Lloyd Ruocco) on the site of the old Scripps Community House (1964)

von Storch: And then what happened?

Munk: We chose the site of the old Community House, half way up the cliffs, where I had spent my first summer back in 1939. Some people thought we were too far from the rest of Scripps, then clustered near the foot of the pier. (Today we are nearer the center of Scripps.) There was concern about the stability of the cliff edge. The University architects hired a soil specialist who quoted Scripps Professor Francis Shepard in concluding that "consistent with other considerations" the building should be placed as far to the east as possible. At a crucial meeting I said "it was"; I had run into Francis on the way to the meeting, and he told me that he had changed his mind.

Judith chose an architect, Lloyd Ruocco, who had built some fine residences, but nothing even close to a laboratory. Judith chose him for his taste, and the fact that his best houses were built when there was not enough money (a situation we were in). At first Lloyd was very reluctant to sign a contract with the University. At our weekly planning meetings he would say, "That's the right way to do it but the University would never buy it." We prevailed (Fig. 11.4). IGPP was built according to plan, on budget, and in time (Fig. 11.5).

---

the sea were absolutely calm you would sea a single reflection of the solar disk. When roughened by even a slight wind, you see millions of glitter points, each designating a facet with the slope required to reflect the Sun into your eyes. At that time the AFOSR Director had asked, "Are you still looking for a photographic plane? You can have our B17 for six weeks if you are ready by August." I accepted. Next day Chip Cox came by asking for suggestions for a thesis. The required images were taken over the Alenuihaha Channel between the islands of Hawaii and Maui. Chip, who as born in Hawaii, was shoeless. I remember him in the transparent bubble of the B17 leaning forward to select a site and triggering the camera with his naked toes. The thesis became one of the most cited papers of the Scripps Institution [43]. Later Chip was elected to the National Academy of Sciences.

Fig. 11.5 Judith and Walter Munk overlooking the Pacific Ocean on the deck of the newly completed IGPP building at Scripps (1964). In 1993 the building was officially named the "Judith and Walter Munk Laboratory"

A serious objection had to do with the choice of redwood for the building material; the University architect worried whether it would be sufficiently permanent (after almost fifty years it is doing just fine). In the early 1970's it was fashionable to equip laboratories with movable walls. Judith found them ugly and expensive, and preferred to allow for rare changes by knocking out a few two-by-fours. Our laboratories were originally partitioned into 2/3 wet space and 1/3 electronics. The walls have now been moved to allow for 1/3 wet space and 2/3 electronics and computing. Among the many objections were the general use of carpets: only the President's office was so authorized. We got away by calling it "acoustic floor covering." An important innovation was to face the laboratories toward an open courtyard where equipment could be assembled and tested in portable laboratories, subsequently trucked and secured on the afterdeck of our vessels. Accordingly all Scripps vessels have screw holes on their fantails four foot on center. I believe we were the first to employ portable laboratories, but this is now a widespread procedure for Oceanographic vessels all over the world (Fig. 6.1).

Fig. 11.6 Ida and Cecil Green with Judith in front of "Spring Stirring," celebrating the Greens' gift of Donal Hord's diorite statue (circa 1964). It has welcomed visitors to IGPP for the past 45 years

von Storch: By now you have been in the building for half a century. What was your subsequent experience?

Munk: We were originally allowed 16,000 of assignable square feet. When we started running out of space, we built upstairs mezzanine offices into the high ceiling laboratories (disguised as "storage closets"). Judith put me into one of these; it shielded me from occasional visitors to the front office, and it took the wind out of the sails of people who felt cheated for not being assigned one of the more elegant upstairs offices.

It is fair to say that people have liked working in the building and the seminar room overhanging the cliffs is very popular. The Canadian oceanographer Bob Stewart sent his architects to La Jolla when he was planning the Institute of Oceanography in Victoria B.C. The English astronomer Fred Hoyle, a frequent visitor, built an Institute of Astrophysics just north of Churchill College in Cambridge. There is an obvious resemblance to IGPP to both laboratories.

Cecil Green had just retired as President of Texas Instruments to take residence in La Jolla. He was an MIT graduate in Electric Engineering. Later he became President of

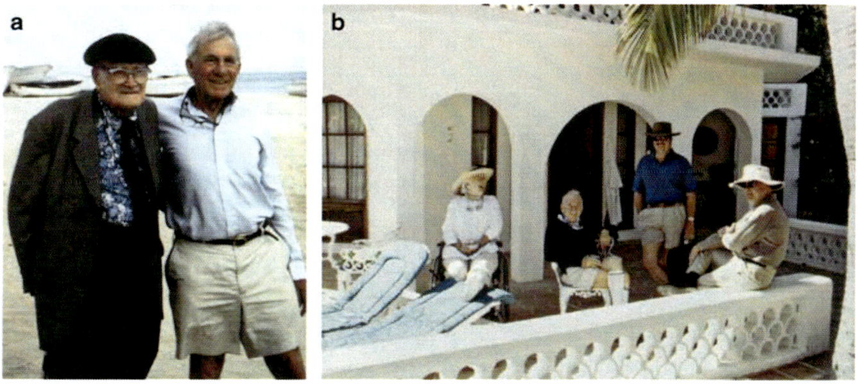

Fig. 11.7 (**a**) Cecil Green and Walter in Cabo San Lucas during the 20th Anniversary Celebration of Project IDA (International Deployment of Accelerometers) (1995) (**b**) Judith, Walter, John Orcutt, and Freeman Gilbert on the patio at Hotel Pamilla following the Celebration

Fig. 11.8 Walter with Jon Berger during the opening ceremonies of the Cecil & Ida Green Piñon Flat Observatory (1980)

Geophysical Services Inc. (GSI), the precursor to Texas Instruments. The 19-story Green Building for geophysical sciences at MIT was dedicated soon after IGPP was built. Cecil came by one evening and said he wanted to give our new laboratory a gift. Judith said at once, "Spring Stirring," her favorite statue by San Diego sculptor Donal Hord. In the earliest back-of-the-envelope sketches Judith had placed "Spring Stirring" at the IGPP entrance. Cecil wanted to make an instant decision, so we drove to Donal's house and climbed over the fence (he had retired) and viewed it by flashlight. Cecil loved it, and so "Spring Stirring" has welcomed our visitors ever since. Donal died two years thereafter, and in his will he had left the statue to Judith (Fig. 11.6).

Cecil and Ida Green were constant supporters of our activities (Figs. 11.7 and 11.8). They supported Freeman Gilbert's global network of digital seismometers (IDA for

International Deployment of Accelerometers). They helped establish the Pinon Flat Observatory above Palm Desert where Jon Berger and others installed a 1 km long laser strain meters. But more important, Cecil felt that we needed more contact with geophysical activities elsewhere, and endowed the Green Scholarship for visitors to IGPP. By now 156 Green Scholars have spent time with us; I would say that they include a majority of workers in our field from all over the world.

Thirty years later we had really run out of space and started planning for the *Roger and Ellen Revelle Laboratory* of the *Cecil and Ida Green Institute of Geophysics and Planetary Physics* (quite a mouthful). Fortunately there was space on the steep hillside across the Biological Grade, on the west side of La Jolla Shores. The challenge was to build the new laboratory in harmony with the old building, but without aping it. We again opted for redwood. By then the University review process has become much more formal, and the plans had to approved by an architectural control committee. The committee was chaired by a San Francisco architect who had been one of Judith's teachers at Harvard. He said, "You cannot use redwood; by the time you start building the price could go sky-high." Some days later Judith said, "He is right of course; why don't we buy the lumber now and let it age for a year." Fortunately we had the building funds at our control, and we hired a wood specialist in Oregon to select the lumber. At the next meeting, the architect said, "I thought I told you not to use redwood, the price might go sky high" to which Judith replied, "You are right, so we bought it already." As it turned out, the price of redwood rose precipitously. We had bought somewhat more than required, and sold the balance at a profit.

Klaus, we spent yesterday afternoon walking through the two buildings. Perhaps you would like to comment?

Hasselmann: Well, for me it was quite a nostalgic experience wandering through the beautiful building in which I had worked in the early sixties, just after it had been built, admiring again the elegant seminar room with its distractive view over the Pacific, and gazing out on to the same view from the spacious window of my old office next door. But I must say, the new building on the adjacent hillside is equally harmonious. Although using the same architectural elements, it has its own distinctive style, using the topology of the hillside, with its many trees and spaces for small patios to create a relaxing Japanese atmosphere.

Munk: One challenge for the new building was to provide for a safe connection with Nierenberg Hall and other Scripps buildings on the east side of La Jolla Shores Drive. Judith and I had spent three months in Tasmania (Fig. 8.5) where we fallen in love with a cable-stayed bridge. So we included a cable-stayed footbridge in the design of the new laboratory. At the time there were no such bridges in Western North America. By luck, UCSD's Dean of the Jacobs School of Engineering, Frieder Seible was an expert on cable-stayed bridges. He said, "Let me build it." In a recent international collection of the best 200 bridges, ours was included.

Fig. 11.9 The Revelle Laboratories of IGPP (designed by Fred Liebhardt) with cable-stayed bridge (designed by Frieder Seible) in background (1993). Judith and Walter fell in love with a cable-stayed bridge in Tasmania and were pleased to discover that UCSD Professor Frieder Seible was anxious to design one. The bridge provided a much needed safe pedestrian access to the Scripps east campus. The bridge has since been included among the world's best 200 bridges

When the old building was named the "Judith and Walter Munk Laboratory" I chose to move across the Grade into the "Roger and Ellen Revelle Laboratory." And there I am today (Fig. 11.9).

Hasselmann: From our discussion it is clear that you took an extraordinary interest in the physical design of the IGPP Laboratories. It is unusual but must be very gratifying for the director of a new institute to become so intimately involved in the details of the building. But, you were, of course, very fortunate in being married to a gifted architect. However, let us turn to a new topic. On a number of occasions you have referred to your relations with the U.S. Navy, and ONR in particular. How did this come about?

Chapter 12

Navy

Munk: It was Roger again. Cdr. Revelle was one of the founders of the Office of Naval Research (ONR), and he saw to it that the Project Officers were good scientists. Among them was John Knauss who later became the Administrator of NOAA, Art Maxwell who went to Woods Hole, and many others.

von Storch: You are speaking of a period long before the establishment of the National Science Foundation. Were the relations with the funding agency different from what they are now?

Munk: Very different. I would describe the relations then as a *partnership* between University and ONR oceanographers. ONR project officers were partners in sharing the excitement of discoveries. Sometimes they would go to sea with us. I recall one occasion when we had lost some vital gear overboard and I needed some guidance of how to proceed, so I telephoned my project officer late at night at home. Continuing financial support depended more on achievements in the past than proposals for the future. Of course, there were only a few of us, and the development of personal trust came naturally. I am very fortunate that the formation of ONR overlapped with my career.

von Storch: I would be concerned that this kind of cozy relation you describe is subject to mis-use.

Munk: It is, and there is no easy answer. I personally much prefer to be subject to the judgment of powerful project officers than to a majority committee vote. With time, as the ratio of research demand to support supply increased, it required near-unanimity to have proposals approved by NSF. This meant that daring proposals with significant chance of failure didn't make it. That is bad for all concerned, bad for the investigator, bad for the funding agency, and bad for oceanography. I feel very strongly about this. Years later Judith and I gave our Kyoto Prize money to Scripps for the benefit of "students engaged in daring research with significant probability of failure," with the provision that the decision is to be made personally by the Director (Fig. 12.1). And we added that if the director does not have a stomach for taking chances, then "so help us God."

H. von Storch, K. Hasselmann, *Seventy Years of Exploration in Oceanography*
DOI 10.1007/978-3-642-12087-9,

Fig. 12.1 Emperor Akihito and Empress Kojun welcoming Judith and Walter on the occasion of the Kyoto Prize (1999)

Fig. 12.2 (**a**) President Ronald Reagan awarding the National Medal of Science to Walter (1984). (**b**) Vice-President George H.W. Bush congratulating Walter (1984)

von Storch: You are getting off the subject.

Munk: Sorry. I want to add that the ONR model led to a cadre of civilian oceanographers who were informed about Navy problems, who cared, and who could offer *meaningful* advice. In times of crisis they were instantly available. When an H-bomb was accidentally dropped off the coast of Spain, the *RV Alvin* was on its way almost at once. I am afraid that much of this post-war spirit has evaporated over the years.

Hasselmann: We have spoken previously of your Navy-related work. Can you try to summarize it?

Munk: There was a progression from wave prediction for amphibious landings to participation in Bikini and Ivy Mike nuclear tests. As part of my fifty years of membership in JASON (a small group of scientists and engineers advising the Department of Defense) we contributed to the development of *non-sound* (we called it un-sound) methods of Anti-Submarine Warfare. We have spoken of Ocean Acoustic Tomography; that has a bearing on Navy problems.

von Storch: Did this interfere with your University career of teaching and research?

Munk: Not at all! To the contrary, I owe much of what I have been able to accomplish to this partnership with the Navy, and with ONR in particular. And I have attempted to recognize this debt. When I received the National Medal of Science from President Reagan in 1985 (Fig. 12.2), I invited Adm. Brad Mooney (then Chief of Naval Research) to come with us to the White House. And in 1999 the then Chief of Naval Research, Admiral Paul Gaffney, came to Japan to participate in the Kyoto Prize award.

Hasselmann: I imagine that a significant fraction of the civilian oceanography community will not agree with you. They think it is healthier to maintain a sharp distinction between civilian and military activities in ocean research. At least that is certainly the case in Europe, and particularly in Germany. Probably, this is because in Germany we have a different history regarding the collaboration between science and the military. In America, you have less stigma attached with the past.

Munk: This is the theme of a recent book on "Oceanography and the Cold War" by J. Hamblin.[1] He argues that the close Navy-University cooperation in the post-war era eroded academic freedom, that Oceanographers "lost their innocence." A review by Munk and Day [261a] takes the opposite position.

von Storch: Walter, throughout your professional life, you have been cooperating with the government of the United States, specifically with the military, both Navy and Air Force. Obviously, you have welcomed this cooperation warmly, ranging from direct consultancy work, such as monitoring for possible tsunamis related to H bomb tests, to more advisory capacities such as in JASON. For somebody who

[1] Hamblin J.D.: *Oceanographers and the Cold War: Disciples of Marine Science*. University of Washington Press, Seattle, WA (2005). See also Rainger R.: Science at the Crossroads: The Navy, Bikini Atoll, and American Oceanography in the 1940's. *Historical Studies in the Physical and Biological Sciences* **30**: 349–371 (2000) and Rainger R.: Patronage and Science: Roger Revelle, the U.S. Navy and Oceanography at the Scripps Institution. *Earth Sciences History* **19**(1): 58–89 (2000).

luckily escaped the Nazis' terror and brainwashing, with a global outreach, who was an active wheel in the war on fascism, this is understandable. Would you mind explaining to us how you see the military dimension of oceanography in general? Is the U.S. military, as compared to, say, the Russian, South African or Japanese military something special? Do events like My Lai or Abu Guraibh have any bearing on these questions?

Munk: People do terrible things in wartime. No country is excluded. But ONR is very special. You will recall that Roger Revelle, John Knauss, Gordon Lill and other oceanographers had a big hand in starting ONR. These were our colleagues. So I have not associated working for ONR with the atrocities you mention.

von Storch: Your membership in JASON brought you rather close to political decision makers in the United States, right? Are you allowed to tell a bit more about JASON, its history and its mode of operation?

Munk: My association with JASON goes back 48 years. It has played a significant role in some national affairs like the test-ban treaty. JASON wrote an early (I think 1979) and influential report on climate change which was prominently mentioned in a recent BBC program on the history of climate change science. I am proud to be a member. JASON was started by some very distinguished people, Hans Bethe, Lyman Spitzer, John Wheeler, Murph Goldberger, Charlie Townes. Not unlike ONR, the incentive for starting JASON was to keep alive a remarkable wartime collaboration between the military and some members of the University science community. And its membership today is distinguished. I don't think there is another group like it in the United States.

I was chairman of the UCSD (University of California at San Diego) faculty during the year of student protests against the Vietnam war. When accepting the appointment I made a public statement that my research was supported by ONR and that I would continue working under Navy sponsorship. Others at UCSD kept their relation with the Defense Department to themselves and when subsequently "discovered" they suffered some quite unpleasant indignities, invasions of their laboratory and rude interruptions of their talks. I had many talks with our students about this subject; would it be better if our military had zero relationship with the University community? In summer 1968 when JASON met on the campus of the University of Colorado in Boulder, there was a student protest against the JASON presence on their campus. In fact, during a lunch break some students pushed their way into the classified facility and threw some typewriters on the floor. It turned out that the leader of that student group had applied for admission to the Scripps Institution, so I was spared.

von Storch: In these days, science is again on the main stage of political deliberations and public attention – climate change. People begin to speak of "war on climate change." Many scientists have become advocates for changes in social organizations and lifestyles. Some even exaggerate in order to increase the perceived urgency of the matter. You were not engaged in these debates, at least not publicly – could you comment on this?

Munk: I am glad you raised this issue. I have worked on some special aspects of climate change, especially sea level change [48, 62, 64, 68, 72, 94, 98, 198, 242, 244, 246, 257]. (i) It is clear that sea level has been rising since long before the industrial revolution. (ii) 20th century rise is at the rate of about 15 cm/century with 5 cm/cy from thermal expansion of the oceans and 10 cm/cy from the melting of continental icecaps. (iii) The rate may now be double this, or 30 cm/cy, with a small and shrinking fraction from thermal expansion. (iv) There have been widespread predictions of 1 m by the end of the century. Perfectly possible; in fact coming out of the ice ages sea level rose 100 m in 10,000 years, or by an *average* of 1 m/cy. But I think the prediction skill is very low. I am impressed that in the last ten years there has been a revolutionary change in the underlying understanding of glacial ice dynamics. (v) Ocean decadal variability is large and can reverse the trend for years at a time. This varies from place to place. Colosi and I looked at 100 years of the Honolulu tide gauge and could not pull out a significant trend. Brian Dushaw has summarized the last ten years of Ocean Acoustic Tomography in the northeast Pacific and does not find a significant warming trend [262]. One would hope that the scientists would place a clearer emphasis on the uncertainty of their extrapolations and that the public would pay more attention to these uncertainties.

von Storch: If I may bring in my personal opinion, maybe just to give you an opportunity to rub against it – we see in the climate issue an example where many scientists are beginning to define policy for society. Society is declared to be unable to embark seriously on the purportedly only possible course of action (for avoiding disaster). To convince society – without overruling formally democratic procedures – methods such as exaggeration of risks, downplaying of uncertainties, and other relevant dynamics are employed to emphasize the urgency. Many climate scientists, who actually are advocates of a political-ideological agenda of making a better world, with better people, try to present themselves as "one-handed" scientists, who give an unequivocal analysis and advice. My personal view is that we must remain "two-handed" professionals, who are admitting that there is always a remaining risk of error, even if I consider it with respect to the human origin of most of the ongoing warming as miniscule. Scientists are not politicians. The present situation, with lots of advocate scientists, is damaging the reputation of science as a valuable social institution, who gives advice but not recommendations in politically charged situations. The buzz words are post-normal science and honest brokers. Your experience with JASON was certainly a totally different one than mine – keeping in mind that I am 30 years less old than you, living in a different country. Interestingly, among JASON members are some known of being fierce "skeptics" in the climate change debate.

Munk: I totally agree, Hans, that we must not let our social preferences get mixed up with reporting on science issues. Was not the mixing of these issues the basis of the Oppenheimer tragedy. This has been an issue at JASON. One of our members, who is among the worlds leading scientist, but also among the best-known contrarians, was so bothered by the unanimity of response of the International Panel on Climate Change (IPCC) that he felt driven to a negative position on climate change (he would rather be wrong than be boring). I have some sympathy for his position but I do not agree. I believe there is clear and convincing evidence for a human contribution to climate change by way of increased atmospheric carbon dioxide. But there are strong contributions from other sources, such as the Ocean Decadal Variability, which can reverse the trend for many years at a time. John Colosi and I looked at a century of Honolulu Sea Level [257] and could not detect a trend over and above decadal variability. On the other side of the climate issue, there are so many who do not tolerate any doubts about the prevailing theory. If you express doubts about some conclusion you are declared an enemy of the planet.

Hasselmann: I know your critical view of the so called unanimity of IPCC, Hans, but I really disagree with you and perhaps Walter on this point. I regard IPCC as giving a really balanced consensus view on the climate problem, including the many uncertainties involved. It is true that individual scientists sometimes overstate the problem, but that really cannot be said of IPCC. And it also true that unqualified people (the "climate skeptics") often create unnecessary noise in the debate, which is then happily amplified by the media. There is unfortunately nothing more boring for the media than to repeatedly report on the pending dangers of climate change, which scientists have been warning of using the same words for decades. So the media are happy for any controversial view, regardless of its scientific basis. And the debate is then happily exploited by interest groups. I feel very strongly that scientists (including JASON) should recognize their responsibility in not fuelling interest groups via the media simply to attract some attention.

von Storch: What do you think about the role, or mission, of science in the society? Science is an expensive exercise – what is, or should be, the rationale for a society to engage in such? Is it the promise of more Teflon pans? What is in your opinion the role of science in advising, influencing, or shaping policy? Is this different in different cultures and nations? How have you dealt with this question?

Munk: It makes a big difference. JASON, for example, flourished during the Kennedy era, when we had contact at the highest level of the administration. JASON languished under the George W. Bush administration. We have had a renaissance of activity since the election of President Obama.

von Storch: The media. I consider this social institution a relevant independent check among the different social powers. While many scientists consider journalists as mostly people, who have to faithfully report, maybe in popular formulations, about the truth released by scientists. But, in reality, journalists do not do that, at least good journalists. Instead they try to tell the full "story," of which the scientists account is just one angle. Therefore, many scientists are frustrated about their contacts with the media. I for my part, I may say that I have been treated fairly or professionally by journalists, with the exception of the BBC, who misquoted me once. How were your experiences with the media?

Munk: Earlier in this discussion I have mentioned the front page article in the *Los Angeles Times* (later retracted) that Acoustic Tomography would lead to the death of 750,000 California Gray Whales (the entire population). This led directly to a widespread opposition to our work which has never quite gone away. Except for that case (a very important one in my career) I have been treated more than fairly by the media.

von Storch: In 1945 Vannevar Bush published his report Science – The Endless Frontier, which described that in return for the privilege of receiving public support, the researcher was obligated to "produce and share knowledge freely to benefit – in mostly unspecified and long-term ways – the public good." Specifically he asserted, "The centers of basic research... are the wellsprings of knowledge and understanding. As long as they are vigorous and healthy and their scientists are free to pursue the truth wherever it may lead, there will be a flow of new scientific knowledge to those who can apply it to practical problems in Government, in industry, or elsewhere." And in 1998, more than fifty years later, a committee of the U.S. House has a resolution, "The United States has been operating under a model developed by Vannevar Bush in his 1945 report. It continues to operate under that model with little change. This approach served us very well during the Cold War...."

Do you think that this Vannevar Bush doctrine, if I may call it so, is still adhered to, at least in the U.S., or has it been replaced by a utility-driven philosophy of more management, top-down control and more immediate, predictably useful results? If so, have you observed a change from Vannevar Bush to this utility-driven research.

Munk: I believe that there are two related principles to the Vannevar Bush doctrine: A: in return for public support, the researcher must share his results with the public, and B: scientists are free to pursue the truth wherever it may lead. I would think that scientists everywhere aspire to these ideals. In the last eight years there have been reports of suppression, or at least alterations, of results when they did not suit the sponsor. Our fervent hope is that all this is now over. As for myself, for the last seventy years, working largely under ONR sponsorship, I have never felt the slightest pressure of directing my work in a direction other than the direction I chose. In the mid sixties when we first proposed to measure deep sea tides, a program officer did come back with the comment that the subject of ocean tides had "gone to bed with Victorian mathematicians." As it turned out, the subject has since enjoyed a renaissance.

Chapter 13
Finis

Hasselmann: Perhaps it is time to come to an end with our prolonged discussion. We started with your early work on ocean surface waves, and its extension to ever lower frequencies. More recently you referred to the use of sound waves for measuring the warming of ocean basins. What are you doing now?

Munk: I flunked retirement. Since 1985 I have had the support of a "Secretary of the Navy Chair in Oceanography" which has given me complete freedom to pursue subjects of my choice (Fig. 13.1).

Fig. 13.1 Navy Chairs with Admiral Carr, Chief of Naval Research (2009). (*Left* to *right*: Michael Gregg, Rob Holman, Frank Herr, Admiral Carr, Tom Sanford, Walter Munk, Tommy Dickey, Arthur Baggeroer, William Kuperman. Not shown: John Orcutt and Robert Weller)

DOI 10.1007/978-3-642-12087-9, © Springer 2010

Fig. 13.2 IUGG (International Union of Geodesy and Geophysics) Symposia in Vienna, Austria (1991). Walter took a bus load of colleagues to Schloss Ziegersberg, his cousin's summer residence south of Vienna. They are pictured having lunch just outside the 12th century Ruine Ziegersberg

Fig. 13.3 Celebrating Walter's 90th birthday hosted by three former students (Paola Rizzoli, Peter Worcester, and Jim Cairns) in Perugia, Italy (2007). (Peter and his wife, Donna, *first* and *third from the left*; Mary Revelle-Paci and her son, Stefano, *second* and *fourth from the right*)

Fig. 13.4 Klaus Hasselmann, Hans von Storch, and Walter having a drink at the end of "The Discussion" (2008)

v. Storch: I can see where you get your prejudices.

Munk: Guilty!

I like to think back of the occasional IUGG meetings held at different international sites (Figs. 13.2 and 13.3); these gave us the opportunity to review our work with our colleagues abroad.

For the last two years I have spent most of my time on surface waves of high frequency, short gravity waves and capillaries. Fifty years ago I worked with Chip Cox (his dissertation) on using photographs of sun glitter to infer the probability of sea surface slopes [54]. If the sea were glassy calm you would see a single image of the sun. Even at slight winds this is spread to a pattern of thousands of glitter points. Each glitter point represents a facet (possibly quite small) with the slope appropriate for reflecting the sun into the camera lens. The outer edges of the glitter pattern define the largest slopes and these increase with increasing wind speed. In 1956 we derived the slope statistics from 28 images taken by B-17 aircraft flying over the Alenuihaha Channel between the islands of Maui and Hawaii. In 2006 Bréon and Heriot derived the slope statistics from 8 million glitter images taken from satellite and distributed globally. I have been trying to understand the Bréon and Heriot data [260].

Fig. 13.5 Celebrating Walter's 65th birthday, a surprise party arranged by Chris Garrett and Carl Wunsch (19 October 1982)

There is an enormous literature on the wave height (or energy) spectrum. This is associated with comparatively long waves, say 10 m and longer; they are the ones that make you seasick. Little is known about waves, shorter than 1 m (say). It is the shorter waves that make up the slope spectrum. These short waves are perhaps more important from an oceanographic point of view than the much studied longer components of the ocean wave spectrum. They are responsible for much of the wind drag. It is the wind drag that keeps the ocean from being a stagnant smelly pool.

Hasselmann: So you think that the satellite glitter images could lead to a better understanding and more accurate measurements of wind drag?

Munk: Yes. Time for a drink! (Figs. 13.4 and 13.5)

Chapter 14
Epilogue

Of course, the story does not end with a drink. Walter ended the talk he gave on the occasion of his 90th birthday symposium (mentioned in our preface) with the remark, "I do not fully understand all of this – but I will!" And indeed he did, as documented by a number of papers in press or published [260–263] on this and related subjects since this book went to press.

It was therefore no surprise that Walter's continuous creative productivity over 70 years received yet another recognition in January 2010 with the award of the prestigious Crafoord Prize by the Royal Swedish Academy of Sciences. Together with the Kyoto prize, which Walter received in 1999, the Crafoord Prize represents one of the most prestigious awards in science after the Nobel Prize.

Hans v. Storch and Klaus Hasselmann

H. von Storch, K. Hasselmann, *Seventy Years of Exploration in Oceanography*
DOI 10.1007/978-3-642-12087-9, © Springer 2010

Appendix A: Curriculum Vitae

Education

California Institute of Technology, B.S. (1939), M.S. (1940)

Scripps Institution of Oceanography, University of California, Ph.D. (1947)

Positions

Professor (Assistant, Associate, Full) Scripps Institution of Oceanography (1947–Present)

Director, La Jolla Laboratory of the Institute of Geophysics and Planetary Physics, Scripps Institution of Oceanography (1959–1982)

Secretary of the Navy Chair in Oceanography (1985–Present)

Honors

Guggenheim Fellow, Oslo University (1948), Cambridge University (1955 & 1962)

Arthur L. Day Medal, American Geological Society (1965)

Harald Sverdrup Gold Medal, American Meteorological Society (1966)

Alumni Distinguished Service Award, California Institute of Technology (1966)

Gold Medal, Royal Astronomical Society (1968)

California Scientist of the Year, California Museum of Sciences and Industry (1969)

Josiah Willard Gibbs Lecturer, American Mathematical Society (1970)

Lockheed Martin Award (1970)

Doctor Philosophiae Honoris Causa, University of Bergen, Norway (1975)

Agassiz Medal, National Academy of Sciences (1976)

Maurice Ewing Medal, American Geophysical Union and U.S. Navy (1976)

Senior Queen's Fellow, Australia (1978)

The Captain Robert Dexter Conrad Award, Department of the Navy (1978)

UCSD Alumnus of the Year (1978 & 1992)

Fulbright Fellow, U.K. (1981–82)

National Medal of Science (1985)

Doctor Philosophiae Honoris Causa, University of Cambridge, England (1986)

William Bowie Medal, American Geophysical Union (1989)

Walter Munk Award, Office of Naval Research & The Oceanography Society (1993)

Vetlesen Prize, Columbia University (1993)

Presidential Award of the New York Academy of Sciences (1993)

Doctor Philosophiae Honoris Causa, University Crete, Greece (1996)

Rolex Lifetime Achievement Award, Singapore (1997)

Kyoto Prize, Inamori Foundation, Japan (1999)

Albert A. Michelson Award, Navy League of the United States (2001)

Prince Albert I Medal, International Association for the Physical Sciences of the Oceans (IAPSO) (2001)

Lifetime Achievement Award, The American Society of Mechanical Engineers (2007)

Spirit of Surfing Award, The Groundswell Society (2007)

Crafoord Prize, The Royal Swedish Academy of Sciences (2010)

Memberships

Fellow, American Geophysical Union (1943)

Member, National Academy of Sciences (1956)

Chairman, Geophysics Section (1975–78)

Chairman, Ocean Studies Board (1985–88)

Fellow, American Academy of Arts and Sciences (1957)

Member, American Philosophical Society (1965)

Elected Fellow, Geological Society of America (1966)

Elected Fellow and Honorary Member, American Meteorological Society (1966)

Foreign Fellow, The Royal Society, London (1976)

Associate, Royal Astronomical Society (1983)

Foreign Member, Russian Academy of Sciences (1994)

Honorary Member, Royal Meteorological Society (1998)

Honorary Member, European Geophysical Society (2000)

Foreign Member, European Academy of Sciences (2004)

Honorary Fellow, Acoustical Society of America (2004)

Member, Deutsche Akademie der Naturforscher Leopoldina

Member, European Academy of Science

Fellow, American Association for the Advancement of Science

Fellow, Acoustical Society of America

Fellow, Marine Technological Society

Appendix B: Bibliography[1]

1941

1 Internal waves in the Gulf of California. *J. Mar. Res.* **4**(1): 81–91

1946

2 H.U. Sverdrup and W.H. Munk. Empirical and theoretical relations between wind, sea and swell. *Trans. Am. Geophys. Union* **27**(6): 823–827

3 H.U. Sverdrup and W.H. Munk. Theoretical and empirical relations in forecasting breakers and surf. *Trans. Am. Geophys. Union* **27**(6): 828–836

1947

4 W.H. Munk and M.A. Traylor. Refraction of ocean waves: a process linking underwater topography to beach erosion. *J. Geol.* **55**(1): 1–26

5 Tracking storms by forerunners of swell. *J. Meteorol.* **4**(2): 45–57

6 Increase in the period of waves traveling over large distances; with application to tsunamis, swell, and seismic surface waves. *Trans. Am. Geophys. Union* **28**(2): 198–217

7 H. Sverdrup and W. Munk. *Wind, sea, and swell: theory of relations for forecasting*. U.S. Hydrographic Office Tech Rpt. No. 1, H.O. Pub. **601**: 1–44

8 Effect of Earth's rotation upon convection cells. *Ann. N.Y. Acad. Sci.* **48**(8): 815–820

9 A critical wind speed for air sea boundary processes. *J. Mar. Res.* **6**(3): 203–218

[1] A reference such as [149b] refers to the second Non-Research Paper published following Research Paper [149] and prior to Research Paper [150]. This distinction was introduced by Munk early in his career to separate papers with new results from those in response to published comments and other discussions. The definitions of "Research Papers" and "Non-Research Papers" are rather loose. It will be noticed that the fraction of papers in the latter category increases with the age of the author. The papers are organized according to the year of publication; numbers are generally in order of acceptance for publication, and therefore not sequential. An exception is [#169] which was written in 1952 but remained classified until 1981. Paper [#115] (written in 1969) is still classified.

1948

10 Wave action on structures. *Petrol. Technol.* **11**(2) Art. TP **2322**: 1–18

11 W.H. Munk, H. Iglesias, and T. Folsom. An instrument for recording ultra-low-frequency ocean waves. *Rev. Sci. Instrum.* **19**(10): 654–658

12 W.H. Munk and M. Sargent. Adjustment of Bikini Atoll to ocean waves. *Trans. Am. Geophys. Union* **29**(6): 855–60

14 W.H. Munk and E.R. Anderson. Notes on a theory of the thermocline. *J. Mar. Res.* Sverdrup Anniv. Vol. **7**(3): 276–295

18 R. Revelle and W. Munk. Harald Ulrik Sverdrup – an appreciation. *J. Mar. Res.* Sverdrup Anniv. Vol. **7**(3): 127–131

1949

13 W.H. Munk, G.C. Ewing, and R.R. Revelle. Diffusion in Bikini Lagoon. *Trans. Am. Geophys. Union* **30**(1): 59–66

15 Note on period increase of waves. *Bull. Seism. Soc. Am.* **39**(1): 41–45

16 The solitary wave theory and its application to surf problems. *Ann. N.Y. Acad. Sci.* **51**(3): 376–424

17 J.A. Putnnan, W.H. Munk, and M.A. Traylor. The prediction of longshore currents. *Trans. Am. Geophys. Union* **30**(3): 337–345

19 Surf beats. *Trans. Am. Geophys. Union* **30**(6): 849–854

1950

20 On the wind-driven ocean circulation. *J. Meteorol.* **7**(2): 79–93

21 W.H. Munk and R.L. Miller. Variations in the Earth's angular velocity resulting from fluctuations in atmospheric oceanic circulation. *Tellus* **2**(2): 93–101

22 W.H. Munk and G.F. Carrier. The wind-driven circulation in ocean basins of various shapes. *Tellus* **2**(3): 158–67

23 W.H. Munk, G.W. Groves, and G.F. Carrier. Note on the dynamics of the Gulf Stream. *J. Mar. Res.* **9**(3): 218–38

25 Oceanography, ocean currents, waves. *Collier's Encyclopedia*: 100-04, 170–71, 398

1951

24 W. Munk and E. Palmén. Notes on the dynamics of the Antarctic circumpolar current. *Tellus* **3**(1): 53–55

27 Y. Mintz and W. Munk. The effect of winds and tides on the length of day. *Tellus* **3**(3): 117–121

28 W.H. Munk and R.S. Arthur. Forecasting ocean waves. Compendium of Meteorol. Am. Meteorol. Soc.: 1082–89

29 Ocean waves as a meteorological tool. Compendium of Meteorol. Am. Meteorol. Soc.: 1090–1100

1952

26 Origin and generation of waves. In: *Proc. First Conference Coastal Eng.* 14 Oct. 1950, Long Beach. Reprinted U.S.N. Civ. Eng. Corps Bull. **6**(5): 133–36

30 W. Munk and R. Revelle. Sea level and the rotation of the Earth. *Am. J. Sci.* **250**: 829–33

31 W. Munk and G. Riley. The absorption of nutrients by aquatic plants. *J. Mar. Res.* **11**(2): 215–40

32 R.S. Arthur, W.H. Munk, and J.D. Isaac's. The direct construction of wave rays. *Trans. Am. Geophys. Union* **33**(6): 855–865

33 W. Munk and R. Revelle. On the geophysical interpretation of irregularities in the rotation of the Earth. *Mon. Not. R. Astron. Soc., Geophys. Suppl.* **6**(6): 331–47

34 W. Munk and G. Groves. The effect of winds and ocean currents on the annual variation in latitude. *J. Meteorol.* **9**(6): 385–96

35 W.H. Munk and R.S. Arthur. Wave intensity along a refracted ray. Natl. Bureau Standards Circular, Gravity Waves **521**: 95–108

37 G.F. Carrier and W.H. Munk. On the diffusion of tides into permeable rock. In: *Wave Motion and Vibration Theory, Proc., Symp. in Appl. Math.* 169 pp. **5**: 89–96, McGraw Hill

39 The roughness of the air-sea boundary. Geophys. Res. Pap. No. 19, Internal. Symp. on Atmospheric Turbulence in the Boundary Layer, MIT, 4–8 Jun. 1951, Geophys. Res. Directorate, AF Cambridge Res. Ctr. Tech. Rpt. 53-9: 141–144

41 Air-sea boundary processes. Oceanographic Instrumentation, Natl. Acad. Sci., Natl. Res. Counc., Pub. 309: 159–161, 210–213

169 W. Bascom, W. Munk, and W. Van Dorn. Barometric and water-surface waves produced by Mike Shot. Operation IVY, Joint Task Force 132, WT-635, 1952 (declassified 1981)

1953

36 Small tsunami waves reaching California from Japanese earthquake of 4 March 1952. *Bull. Seism. Soc.* **43**(3): 219–222

1954

12b W.H. Munk and M.C. Sargent. Adjustment of Bikini Atoll to ocean waves. Geol. Survey Professional Pap. 260C: 275–280

38 E. Gossard and W. Munk. On gravity waves in the atmosphere. *J. Meteorol.* **11**(4): 259–269

40 The delayed hot-water problem. *J. Appl. Mech.* **21**(2): 193

42 Y. Mintz and W. Munk. The effect of winds and bodily tides on the annual variation in the length of day. *Mon. Not. R. Astron. Soc., Geophys. Suppl.* **6**(9): 566–578

43 C. Cox and W. Murk. Measurement of the roughness of the sea surface from photographs of the Sun's glitter. *J. Opt. Soc. Am.* **44**(11): 838–850

44 C. Cox and W. Munk. Statistics of the sea surface derived from sun glitter. *J. Mar. Res.* **13**(2): 198–227

1955

45 C. Cox and W. Munk. Some problems in optical oceanography. *J. Mar. Res.* **14**(1): 63–78

46 The circulation of the oceans. *Sci. Am.* **193**(3): 96–104

47 Wind stress on water: a hypothesis. *Q. J. Roy. Meteorol. Soc.* **81**(349): 320–332

48 J. Pattullo, W. Munk, R. Revelle, E. Strong. The seasonal oscillation in sea level. *J. Mar. Res.* **14**(1): 88–155

49 R. Revelle and W. Munk. Evidence from the rotation of the Earth. UGGI Bull., Extrait des Annales de Geophysique **11**(1): 104–108

50 E. Gossard and W. Murk. Gravity waves in the atmosphere: correspondence. *Q. J. Roy. Meteorol. Soc.* **81**(349): 484–487

53 High-frequency spectrum of ocean waves. WHOI Convocation, *J. Mar. Res.* **14**(4): 302–314

1956

48a Geophysical discussion. *Roy. Astron. Soc.*, reprinted from *The Observatory* **76**: 56–57

51 W. Munk, F. Snodgrass, and G. Carrier. Edge waves on the continental shelf. *Science* **123**(3187): 127–132

52 Polar wandering: a marathon of errors. *Nature* **177**: 551–554

54 C. Cox and W. Munk. Slopes of the sea surface deduced from photographs of sun glitter. Scripps Institution of Oceanography Bull. **6**(9): 401–488

1957

55 Waves of the Sea. *Encyclopedia Britannica*

56 Comments on review by C. L. Bretschneider. *Trans. Am. Geophys. Union* **38**(5): 778–781

57 W.H. Munk and F.E. Snodgrass. Measurements of southern swell at Guadalupe Island. *Deep-Sea Res.* **4**: 272–286

59 W.H. Munk, M.J. Tucker, and F.E. Snodgrass. Remarks on the ocean wave spectrum. In: Symposium on Naval Hydrodynamics, Pub. No. 515, Chap. **3**: 45–60

65a Harald Ulrik Sverdrup, in memoriam. *Deep-Sea Res.* **4**(4): 289–290

1958

58 F. Snodgrass, W. Munk, and M.J. Tucker. Offshore recording of low-frequency ocean waves. *Trans. Am. Geophys. Union* **39**(1): 114–120

60 W. Elsasser and W. Munk. Geomagnetic drift and the rotation of the Earth. Gutenberg Volume, "Contributions in Geophysics: In Honor of Beno Gutenberg," *Deep-Sea Res.* **17**: 228–236

61 The seasonal budget of water. In: *Geophysics and the I.G.Y., National Academy Symposium*, ed. H. Odishaw *et al.*, Geophys. Monogr. No. 2, pp. 175–176. Wash., D.C.: Am. Geophys. Union

62 W. Munk and R. Haubrich. The annual pole tide. *Nature* **182**: 42

63 Waves of the sea. *The Book of Popular Science*, pp. 361–368

64 Remarks concerning the present position of the pole. *Geophysica* **6**(34): 335–355

71 G.W. Groves and W.H. Munk. A note on tidal friction. *J. Mar. Res.*, Tommy Thompson Vol. **17**: 199–214

1959

65 B. Haurwitz, H. Stommel, W. Munk. On the thermal unrest in the ocean. In: *The Rossby Memorial Volume,* pp. 74–94, New York. The Rockefeller Institute Press

66 H. Benioff, J.C. Harrison, L. LaCoste, W.H. Munk, and L.B. Slichter. Searching for the Earth's free oscillations. *J. Geophys. Res.* **64**(9): 1334–1337

67 W.H. Munk, F.E. Snodgrass, and M.J. Tucker. Spectra of low-frequency ocean waves. Scripps Institution of Oceanography Bull. **7**: 283–362

68 R. Haubrich and W. Munk. The pole tide. *J. Geophys. Res.* **64**(12): 2373–2388

1960

69 Smoothing and persistence. *J. Meteorol.* **17**(1): 92–93

70 W.H. Munk and G.J.F. MacDonald. Continentality and the gravitational field of the Earth. *J. Geophys. Res.* **65**(7): 2169–2172

72 W. Munk and G.J.F. MacDonald. *The Rotation of the Earth: A Geophysical Discussion*. Cambridge University Press, 323 pp

1961

73 W.H. Munk and E.M. Hassan. Atmospheric Excitation of the Earth's wobble. *Geophys. J. R. Astron. Soc.*, Jeffreys Jubilee Vol. **4**: 339–358

74 J. Miles and W. Munk. Harbor paradox. In: *Proc. Am. Soc. Civ. Eng., J. Waterways and Harbors Div.* **87**(WW3): 111–130

75 W. Munk and H. Cepeda. Concerning a remarkably sharp peak in the sea level spectrum at Acapulco. *Geophysicia Int.* **1**(3): 45–54

1962

76 D.L. Inman, W.H. Munk, and M.A. Balay. Spectra of low-frequency ocean waves along the Argentine shelf. *Deep Sea Res.* **8**: 155–164

77 Introduction to *The Tides*, by George Howard Darwin, pp. ix–xv, San Francisco: H. Freeman and Company

78 F.E. Snodgrass, W.H. Munk, and G.R. Miller. Long-period waves over California's continental borderland. Part I. Background spectra. *J. Mar. Res.* **20**(1): 3–30

79 G.R. Miller, W.H. Munk, and F.E. Snodgrass. Long-period waves over California's continental borderland. Part II. Tsunamis. *J. Mar. Res.* **20**(1): 31–41

80 W.H. Munk, G.R. Miller, and F.E. Snodgrass. Long-period waves over Califomia's continental borderland. Part III. The decay of tsunamis and the dissipation of tidal energy. *J. Mar. Res.* **20**(2): 119–120

81 Long ocean waves. In: *The Sea*, ed. M.N. Hill, Vol. **1**, and Sect. 5, Chap. 18, pp. 647–663, London: Interscience

108 W. Munk, R. Haubrich, and G. Backus. Array studies of background noise. In: *Proc. of the Colloquium on Detection of Underground Nuclear Explosions, Acoustic and Seismic Lab.*, Inst. of Sci. and Tech., Univ. of Michigan, Ann Arbor, Aug. 1962: 95–124

1963

82 R.A. Haubrich, W.H. Munk, F.E. Snodgrass. Comparative spectra of microseisms and swell. *Bull. Seism. Soc. Am.* **53**(1): 27–37

83 K. Hasselmann, W. Munk, and G.J.F. MacDonald. Bispectra of ocean waves. In: *Time Series Analysis*, ed. M. Rosenblatt. Chap. 8, pp. 125–139, John Wiley and Sons

84 W.H. Munk, Moderator. Panel discussion, "Ocean Wave Spectra." In: *Proc. of Conf., 1–4 May 1961, Easton, Md.*, pp. 333–345, Prentice-Hall: arranged by NAS

85 W. Munk, G.R. Miller, F.E. Snodgrass, N.F. Barber. Directional recording of swell from distant storms. *Philos. Trans. R. Soc. London* A, No. 1062, 255: 505–584

86 W.H. Munk and E.C. Bullard. Patching the long-wave spectrum across the tides. *J. Geophys. Res.* **68**(12): 3627–34

87 Some comments regarding diffusion and absorption of tsunamis. In: *Proc. Tsunamis Meetings*, 10th Pacific Science Congress, Honolulu, 1961, IUGG: 53–72

1964

88 W.H. Munk and D. Davies. The relationship between core accretion and the rotation of the Earth. In: *Isotopic and Cosmic Chemistry*, ed. H. Craig *et al.*, Urey Volume, pp. 341–346, Amsterdam: North Holland

89 W. Munk, F. Snodgrass, and F. Gilbert. Long waves on the continental shelf: an experiment to separate trapped and leaky modes. *J. Fluid Mech.* **20**(4): 529–554

90 W. Munk and K. Hasselmann. Super-resolution of tides. In: *Studies on Oceanography*, ed. K. Yoshida, Hidaka Volume, pp. 339–344, Tokyo. Reprinted Univ. of Wash. Press, 1965

91 E.C. Bullard, F.E. Oglebay, W.H. Munk, and G.R. Miller. A user's guide to BOMM. A system of programs for the analysis of time series. Institute of Geophysics and Planetary Physics, University of California, La Jolla, California, 108 pp

91a W.H. Munk and G.J.F. MacDonald. Third presentation of the James B. Macelwane Award, April 12, 1964: citation (awarded to Klaus Hasselmann). *Trans. Am. Geophys. Union* **45**(2): 289–290

1965

92 W.H. Munk, B. Zetler, and G.W. Groves. Tidal Cusps. *Geophys. J.* **10**. 211–219

92a Waves across the Pacific. An expedition to study how waves lose their energy. *Oceanus* **11**(3): 8–12

93 Tides of the Planet Earth. In: *AFOSR 10th Anniversary Summer Scientific Seminar*, Cloudcroft, N. M., pp. 134–137

95 Acceptance of the Day Award. Geol. Soc. Am. Annual Meeting, Nov., 1965, Kansas City, Mo., *Bull. Geol. Soc. Am.* **76**(12): 233–235

1966

94 Coupling of waves, tides and the climatic fluctuation of sea level. In: *Advances in Earth Science*, ed. P.M. Hurley, pp. 185–195, Boston: MIT Press

94a The Nth campus problem. *Bear Facts*, Vol. **4**, No. 8, Iss. 38

96 F.E. Snodgrass, G.W. Groves, K.F. Hasselmann, G.R. Miller, W.H. Munk and W.H. Powers. Propagation of ocean swell across the Pacific. *Philos. Trans. R. Soc. London*, Ser. A, No. 1103, **259**: 431–497

97 W.H. Munk and D.E. Cartwright. Tidal spectroscopy and prediction. *Philos. Trans. R. Soc. London*, Ser. A, No. 1105, **259**: 533–581

98 Variation of the Earth's rotation in historical times. In: *The Earth–Moon System*, ed. B.G. Marsden *et al.* pp. 52–59, Plenum Press

98a Donal Hord Eulogy. Private distribution

99 Review of Stommel's Gulf Stream: a physical and dynamical description. *Bull. Am. Meteorol. Soc.* **47**(5): 424–425

100 Abyssal recipes. *Deep-Sea Res.* **13**: 707–730

102 The abyssal Pacific. Fifth Marchon Lecture, University of Newcastle Upon Tyne, delivered 26 May 1966

104 Tide Prediction. In: *Proc. from the IBM Scientific Computing Symposium on Environmental Sciences*, 14–16 Nov., Thomas J. Watson Research Center, Yorktown Heights, N.Y.: 143–144

1967

101 R. Radok, W. Munk, and J. Isaacs. A note on mid-ocean internal tides. *Deep-Sea Res.* **14**: 121–124

103 Ocean waves and turbulence. Review of "The Dynamics of the Upper Ocean," by O. M. Phillips. *Nature* **213**: 867

105 J. Houtermans, H.E. Suess, and W.H. Munk. Effect of industrial fuel combustion on the carbon-14 level of atmospheric CO_2. In: *Radioactive Dating and Methods of Low-Level Counting, Vienna:* International Atomic Energy Agency: 57–68

106 W.H. Munk and B.D. Zetler. Deep-sea tides: a program. *Science* **158**: 884–886

107 Some lectures on physical oceanography. In: *Mobil Oil Colloquium Lectures in Pure and Applied Science*, No. 13, 104 pp

1968

108a W. Munk and R. Revelle. Thirtieth award of the William Bowie Medal, April 8, 1968 (awarded to Roger Revelle). Acceptance and response. *Trans. Am. Geophys. Union* **49**(2): 431–434

109 W. Munk and D. Moore. Is the Cromwell current driven by equatorial Rossby waves? *J. Fluid Mech.* **33**(2): 241–259

110 W. Munk and N. Phillips. Coherence and band structure of inertial motion in the sea. *Rev. Geophysics* **6**(4): 447–472

112 Once again-tidal friction. *Q. J. R. Astron. Soc.*, Harold Jeffreys Lecture: **9**(4): 352: 75

1969

111 W. Munk and M. Wimbush. A rule of thumb for wave breaking over sloping beaches. *Oceanology* (USSR), Stockmann Memorial Vol. **9**(1): 71–75

113 Deep-sea tides. In: *Progress in Oceanography*, ed. M. Sears, Vol. **5**, pp. 67–69, Pergamon Press

114 D. Cartwright, W. Munk, and B. Zetler. Pelagic tidal measurements: a suggested procedure for analysis. *EOS, Trans. Am. Geophys. Union* **50**(7): 472–477

115 W.H. Munk, C.G. Callan, R. Dashen, J.B. Hartle, J.B. Keller, J. Miles, W.A. Nierenberg, C. Wright, and F. Zachariasen. On some superficial effects from moving sources in a stratified fluid. *J. Defense Research*, Series B: Tactical Warfare: 1(2), Classified

116 W.H. Munk and W.A. Nierenberg. High frequency radar sea return and the Phillips saturation constant. *Nature* **224**(5226): 1285

117 Low frequency tail of the ocean wave spectrum. In: *Sixth Symposium on Naval Hydrodynamics*, 28 Sept. to 4 Oct. 1966, Wash., D.C., ed. R.D. Cooper *et al.*, ONR ACR-136: 475–476

120a Standard wave spectra for open sea structures. In: Oceanographic Studies, La Jolla Bendix Marine Advisors, 6 pp

1970

119 W. Munk, F. Snodgrass, M. Wimbush. Tides off shore: transition from California coastal to deep-sea waters. *Geophys. Fluid Dyn.* **1**(1): 161–235

120 M. Hendershott and W. Munk. Tides. *Ann. Rev. Fluid Mechan.* **2**: 205–224
121 J.B. Keller and W.H. Munk. Internal wave wakes of a body moving in a stratified fluid. *Phys. of Fluids* **13**(6): 1425–1431
122 Tides of the deep sea. *Explorers Journal* **48**(3): 177–183
123 B. Zetler, D. Cartwright, and W. Munk. Tidal constants derived from response admittances. In: *Sixth International Symposium on Earth Tides*, Strasbourg, Sept. 1969, pp. 175–178
131b Deep-sea tides (abstract only). In: *Eighth Symposium Naval Hydrodynamics*, "Hydrodynamics in the Ocean Environment," 24–28 Aug. 1970, Pasadena, ONR-ACR-179: 217

1971

118 M. Wimbush and W. Munk. The benthic boundary layer. In: *The Sea*, ed. A.E. Maxwell, Vol. **4**, Pt. 1, Chap. 19, pp. 731–758, John Wiley and Sons
124 C.J.R. Garrett, and W.H. Munk. The age of the tide and the "Q" of the oceans. *Deep-Sea Res.* **18**: 493–503
125 C.J.R. Garrett and W.H. Munk. Internal wave spectra in the presence of finestructure. *J. Phys. Oceanogr.* **1**(3): 196–202
125a Tribute to Columbus Iselin. *Oceanus*, Columbus Iselin Iss. **16**(2): 44
126 J. Irish, W. Munk, and F. Snodgrass. M_2 amphidrome in the Northeast Pacific. *Geophys. Fluid Dyn.* **2**: 355–360
127 B.S. Gallagher and W.H. Munk. Tides in shallow water: spectroscopy. *Tellus* **23**(4-5): 346–363
129 Concluding remarks at the Joint Oceanographic Assembly. In: *Proc. of the Joint Oceanographic Assembly*, 13–25 Sept. 1970, Toyko, ed. M. Uda, pp. 81–83
129a The benthic boundary (abstract only). In: *Proc. of the Joint Oceanographic Assembly*, 13–25 Sept. 1970, Toyko, ed. M. Uda, p. 138
129b The circulation near the sea floor (abstract only). In: *Proc. of the Joint Oceanographic Assembly*, 13–24 Sept. 1970, Toyko, ed. M. Uda, p. 230

1972

128 Turbulence in a stratified ocean. Lecture notes in Physics. In: *Proc. of the Symposium on Statistical Models and Turbulence*, 15–21 July 1971, La Jolla, ed. M. Rosenblatt *et al.* pp. 455–449
130 C. Garrett and W. Munk. Space-time scales of internal waves. *Geophys. Fluid Dyn.* **2**: 225–264
131 J. Munk and W. Munk. Venice Hologram. In. *Proc. of the Am. Philos. Soc.* **116**(5): 415–442
131a J.F. Asmus, R F. Wuerker, and W. Munk. Lasers and holography in art preservation and restoration. In: *Proc. of NEREM*, 1–3 Nov. 1972, Boston, pp. 172–175
132 C. Garrett and W. Munk. Oceanic mixing by breaking internal waves. *Deep-Sea Res.* **19**: 823–832

134 W.A. Nierenberg and W.H. Munk. Sea spectra and radar scattering. IDA/JASON Note N-817, Contract DAHC15 67 C 011, Log. No. HQ73-25031, 25pp

1973

133 W.H. Munk and C.J.R. Garrett. Internal wave breaking and microstructure (the chicken and the egg). *Boundary-Layer Meteorol.* **4**: 37–45

133a W.H. Munk, J.D. Woods, W.P. Berkemeier, C.G. Little, R.H. Stewart and J. Richter. Working group report II – remote sensing of the ocean. 1972 IUCRM Colloquium, La Jolla. *Boundary-Layer Meteorol.* **5**: 201–209

134a W. Brown, W. Munk, F. Snodgrass, and B. Zetler. Bottom pressures in the MODE region. *MODE Hot Line News*, 37: 1, 4

135 Tides. In: *Oceanography – The Last Frontier*, ed. R.C. Vetter, Chap. 13, pp. 179–187, New York: Basic Books

136 W.H. Munk, F. Snodgrass, and B. Zetler. Oceanography from the bottom up. *Naval Res. Rev.* **26**(9): 18–24

138 J.F. Asmus, C.G. Murphy, and W.H. Munk. Studies on the interaction of laser radiation with an artifacts. In: *Proc. of the SPIE Conference*, 27–29 Aug. 1973, San Diego, Vol. 41, pp. 19–27

144 F. Rizzoli and W. Munk. Alcune considerazioni teoriche sul coefficente di riflessione del sistema parete forata – parete solida. Tr. 61 of Consiglio Nazionale Delle Ricerche, Laboratorio per lo Studio Della Dinamica Delle Grandi Masse, Venezia, Feb., 1973, 23 pp

1974

137 Sound channel in an exponentially stratified ocean, with application to SOFAR. *J. Acoust. Soc. Am.* **55**(2): 220–226

139 G.L. Tyler, C.C. Teague, R.H. Stewart, A.M. Peterson, W.H. Munk and J.W. Joy. Wave directional spectra from synthetic aperture observations of radio scatter. *Deep-Sea Res.* **21**: 989–1016

139a J. Brune and W. Munk. Maurice Ewing Obituary. *Nature* **252**: 429

140 M.C. Hendershott, W.H. Munk, and B.D. Zetler. Ocean tides from SEASAT-A. In: *SEASAT-A Contributions*, Wash., D.C.: NASA, pp. 54–56

143a Deep sea tides (abstract only). In: *Proc. Int'l Conf., Physics of the Seas*, 13–16 Oct. 1971, Centro Internazionale di Fisica Teorica, Trieste, Quad. No. 206, pp. 103–104

143b Vertical diffusion due to internal wave breaking. In: *Proc. Int'l Conf., Physics of the Seas*, 13–16 OCL 1971, Centro Internazionale di Fisica Teorica, Trieste, Quad No. 206, 1 pg

1975

141 C. Garrett and W. Munk. Space-time scales of internal waves: a progress report. *J. Geophys. Res.* **80**(3): 291–297

142 F. Snodgrass, W. Brown, and W. Munk. MODE: IGPP measurements of bottom pressure and temperature. *J. Phys. Oceanogr.* **5**(1): 63–74
143 W. Brown, W. Munk, F. Snodgrass, H. Mofjeld, and B. Zetler. MODE bottom experiment. *J. Phys. Oceanogr.* **5**(1): 75–85
145 B. Zetler, W. Munk, H. Mofjeld, W. Brown, and F. Dormer. MODE tides. *J. Phys. Oceanogr.* **5**(3): 430–441
146 W.H. Munk, H. Stommel, A.S. Sarkisyan, and A.R. Robinson. Where do we go from here? In: *Numerical Models of Ocean Circulation. Proc. of a Symposium*, 17–20 Oct. 1972, Durham, New Hampshire, NAS, Wash., D.C., Chap. 8, Conclusions, pp. 349–360
146a SCOR Working Group 27. An intercomparison of open sea tidal pressure sensors. Rpt. of SCOR Working Group 27: "Tides of the Open Sea," Paris, UNESCO Tech. Paps. in Mar. Sci. 21, 67 pp
147 B.D. Zetler and W.H. Munk. The optimum wiggliness of tidal admittances. *J. Mar. Res.* **33**: 1–13
147a *The Rotation of the Earth: A Geophysical Discussion.* Cambridge University Press, 323 pp. (2nd printing)

1976

148 W.H. Munk and F. Zachariasen. Sound propagation through a fluctuating stratified ocean: theory and observation. *J. Acoust. Soc. Am.* **59**(4): 818–838
148a W.H. Munk and F. Zachariasen. Reply to "comments on 'sound propagation through a fluctuating stratified ocean: theory and observation' ", by J. McCoy, *J. Acoust. Soc. Am.* **60**(5): 1217–1218
149 F. Dyson, W. Munk, and B. Zetler. Interpretation of multipath scintillations Eleuthera to Bermuda in terms of internal waves and tides. *J. Acoust. Soc. Am.* **59**(5): 1121–1133
149a First presentation Maurice Ewing Medal – acceptance and response. 13 April 1976
149b Award of the Agassiz Medal – acceptance and response. 26 April 1976
150 W.H. Munk and R.W. Preisendorfer. Carl Henry Eckart, 1902–1973. In: *Biographical Memoirs*, NAS, Wash. D.C., Vol. **48**, pp. 195–219

1977

151 Huge waves can be freaky – so can huge tankers. *L. A. TIMES*, 26 February 1977, Pt. 2: 5
151b F.N. Spiess, W.H. Munk, J.D. Isaacs, and L.N. Liebemann. Carl Eckart, 1902–1973. In: In Memoriam, University of California, pp. 89–91
151c Graduation at Midland. Unpublished. June, 1977
152 W.H. Munk and G.O. Williams. Acoustic oceanography. *Nature* **267**(5614): 774–778
152a Kidd criticized. LA JOLLA LIGHT, Letters to the Editor, 11 Aug. 1977
153 R. Revelle and W. Munk. The carbon dioxide cycle and the biosphere. In: Studies in Geophysics, Energy and Climate, Part 3, Monitoring and Modeling, Sect. 10, NAS, Wash., D.C., pp. 140–158

154 W.H. Munk and P.F. Worcester. Monitoring the ocean acoustically. In: Science, Technology, and the Modern Navy – Thirtieth Anniversary, 1946–1976, Arlington, Va., (ONR-37): 497–508, 1976. Also appears under the title: Weather and climate under the sea – the Navy's habitat. In: *Science and the Future Navy – A Symposium*, Thirtieth Ann. Vol., ONR, NAS, Wash., D.C, pp. 42–52, 1977

1978

151a Variable ocean structure. In: *Advances in Oceanography*, ed. H. Charnock *et al.*, pp. 209–211, London: Plenum Press

155 Visit to Tsingtao. *EOS, Trans. Am. Geophys. Union* **59**(1) 4–7

155a R.L. Garwin, W.H. Munk, and C. Wunsch. Preliminary report on ocean acoustic monitoring. JASON Rpt. JSN 77-8

155b Edward Bullard – dedication.[2] In: AIP Conf. Proc. No. 46, Topics in Nonlinear Dynamics; A Tribute to Sir Edward Bullard, ed. S. Jorna, pp. v–viii, New York Am. Inst. of Physics

155c W.H. Munk and C. Wunsch. Ocean acoustic tomography: can it work (abstract only)? Invited paper, 95th Meeting Acoust. Soc. Am., Session CC. Underwater Acoustics V: Special Session – Propagation through Fronts, Eddies and Gyres. *J. Acoust. Soc. Am.* **63**(1): S71

155d Third presentation Maurice Ewing Medal to Sir Edward Bullard. *EOS, Trans. Am. Geophys. Union* **59**(8): 794–795

155e Professor Walter Munk at graduate commencement 1978

1979

156 C. Garrett and W. Munk. Internal waves in the ocean. *Ann. Rev. Fluid Mech.* **11**: 339–369

157 W. Munk and C. Wunsch. Ocean acoustic tomography: a scheme for large scale monitoring. *Deep-Sea Res.* **26A**: 123–161

158 S.M. Flatté, R. Dashen, W.H. Munk, K.M. Watson, and F. Zachariasen. *Sound Transmission Through A Fluctuating Ocean.* Cambridge University Press, 299 pp

159 J. Cairns, W. Munk, and C. Winant. On the dynamics of neutrally buoyant capsules; an experimental drop in Lake Tahoe. *Deep-Sea Res.* **26A**: 369–381

[2] "A Tribute to Sir Edward Bullard" was, in fact, written word-for-word by Sir Edward Bullard. I had just completed the presentation of the Maurice Ewing Medal to Sir Edward [155d] when I was asked to write the dedication to Sir Edward of the AIP Conference Proceedings [155b]. I was in no mood to do so, and asked Teddy whether he would enjoy writing about himself what *he* thought ought to be said; it was to remain forever a secret. Teddy wrote a delightful account of his career, full of glimpses such as returning from the War and finding his Cambridge laboratory in shambles: "his first task was to scrub the floor". The manuscript was returned to me by the editors with the above quote and many others deleted. Upon protesting, I was told that Sir Edward would be gravely offended. I persisted. Teddy was so delighted with the dedication that he spilled the beans.

1980

159a Introduction. In: Oceanography in China A Trip Report of the American Oceanography Delegation, NAS, Wash. D.C., 106 pp

160 Affairs of the sea. Prefatory Chapter in *Ann. Rev. Earth Planet. Sci.* **8**: 1–16

161 Horizontal deflection of acoustic paths by mesoscale eddies. *J. Phys. Oceanogr.* **10**(4): 596–604

162 M.G. Brown, W.H. Munk, J.L. Spiesberger, and P.F. Worcester. Long-range acoustic transmission in the Northwest Atlantic. *J. Geophys. Res.* **85**(C5): 2699–2703

162a Ocean acoustic tomography: a method for measuring large-scale variability. In: Proc. NOAA Workshop on Ocean Acoustic Remote Sensing, 21–24 Jan. 1980, Seattle, Wash., vol. 2, Presentations and Working Group Reports, 4-1, 4–10

162b G.J.F. MacDonald (Chairman), *et al.* The oceans as a sink for carbon dioxide. In: The Long-Term Impacts of Increasing Atmospheric Carbon Dioxide Levels, JASON Tech. Rpt. No. JSR-79-04: 65–82

162c Walter Munk Autobiography. McGraw-Hill Modern Scientists and Engineers: 339–340

164 Internal wave spectra at the buoyant and inertial frequencies. *J. Phys. Oceanogr.* **10**(11), 1718–1728

1981

163 Internal waves and small scale processes. In: *Evolution of Physical Oceanography – Scientific Surveys in Honor of Henry Stommel*, B.A. Warren and C. Wunsch (eds.), Chap. 9, pp. 264–91, Boston: MIT Press

165 W. Munk, P. Worcester, and F. Zachariasen. Scattering of sound by internal wave currents: the relation to vertical momentum flux. *J. Phys. Oceanogr.* **2**: 442–454

166 W. Munk, B. Zetler, J. Clark, S. Gill, D. Porter, J. Spiesberger and R. Spindel. Tidal effects on long-range sound transmission. *J. Geophys. Res.* **86**(C7): 6399–6410

1982

167 W. Munk and C. Wunsch. Ocean acoustic tomography: large scale climatic measurements and modeling. *Ocean Modeling* **43**: 14

168 Why is the Earth a poor timekeeper? Book review, *The Earth's Variable Rotation: Geophysical Causes and Consequences*, by Kurt Lambeck. *Geophys. J. R. Astron. Soc.* **69**: 831–835

168a Time series of vorticity, heat and heat flux, and upwelling in ocean gyres from long-range acoustic transmission (abstract). *EOS, Trans. Am. Geophys. Union*, **63**: 362

170 Citation for Bowie Medalist Henry Stommel. *EOS, Trans. Am. Geophys. Union* **63**(28): 522

171 D. Behringer, T. Birdsall, M. Brown, B. Cornuelle, R. Heinmiller, R. Knox, K. Metzger, W. Munk, J. Spiesberger, R. Spindel, D. Webb, P. Worcester, and C. Wunsch. A demonstration of ocean acoustic tomography. *Nature* **299**(5876): 121–125

172 W. Munk and C. Wunsch. Observing the ocean in the 1990's. *Philos. Trans. R. Soc. London.* **307**: 439–64

1983

173 Acoustics and ocean dynamics. In: *Oceanography: The Present and Future*, ed. P. G. Brewer, pp. 109–126, New York: Springer

174 W. Munk and C. Wunsch. Up/down resolution in ocean acoustic tomography. *Deep-Sea Res.* **29**(12A): 1415–1436

175 W. Munk and C. Wunsch. Ocean acoustic tomography: rays and modes. *Rev. Geophys. Space Phys.* **21**(4): 777–793

176 Wave predictability - Short range and long range. In *Proceedings from the Symposium June 8–10, 1983, International Association of Drilling Contractors and Scripps Institution of Oceanography, University of California, La Jolla, CA*

179 W. Munk and C. Wunsch. Time series of vorticity, heat and heat flux in ocean gyres from long-range acoustic transmissions (abstract only). World Climate Research Programme (WCRP) Publication Series No. 1, 329–330

1984

176a W. A. de N. Unpublished talk given at William A. Nierenberg's 65th Birthday Celebration, February 13, 1984, Scripps Institution of Oceanography, University of California, La Jolla, California

177 An observing net for oceanic prediction. Invited paper presented at the La Jolla Institute Workshop, held at Scripps Institution of Oceanography, University of California, La Jolla, California, Feb. 1983. *Am. Instit. Phys.* 505

178 R. Dashen and W. Munk. Three models of global ocean noise. *J. Acoust. Soc. Am.* **76**: 540–554, August, 1984

180 A Celebration in Geophysics and Oceanography – 1982, In Honor of Walter Munk on his 65th Birthday, October 19, 1982 at Scripps Institution of Oceanography, University of Califomia, La Jolla, Califomia. SIO Ref. Series 84–85

1985

181 B. Cornuelle, C. Wunsch, D. Behringer, T. Birdsall, M. Brown, R. Heinmiller, R. Knox, K. Metzger, W. Munk, J. Spiesberger, R. Spindel, D. Webb, and P. Worcester. Tomographic Maps of the Ocean Mesoscale – 1: Pure Acoustics. *J. Phys. Oceanogr.* **15**: 133–152

182 W. Munk and C. Wunsch. Biases and caustics in long-range acoustic tomography. *Deep-Sea Res.* **32**: 1317–1346

183 G. Groves and W. Munk.[3] A decompression gauge for divers. Scripps Institution of Oceanography, University of Califomia, La Jolla, California. SIO Ref. Series 53–64 (1953)

183a Tribute to Frank Snodgrass. *EOS, Trans. Am. Geophys. Union*. **66**(28): 753

1987

184 Acoustic monitoring of ocean gyres. Invited paper presented at IUTAM Symposium on Fluid Mechanics in the Spirit of G.I. Taylor, held at University of Cambridge, Cambridge, England, 24–28 Mar. 1986. *J. Fluid Mech.* **173**: 43–53

184a Introductory remarks to Ocean Principals, Washington, DC, 18 February 1987

185 W. Munk and C. Wunsch. Bias in Acoustic Travel Time through an Ocean with Adiabatic Range-Dependence. *Geophys. Astrophys. Fluid Dynamics* **39**: 1–24

186 W. Munk, P. Scully-Power, and F. Zachariasen. Ships from Space. The Bakerian Lecture, 1986. *Proc. R. Soc. London* **412**: 231–254

1988

187 B.M. Howe and W.H. Munk. Deep sea moorings in a tidal current. *Deep-Sea Research* **35**: 111–119

187a Citation for Gordon Hamilton. Invited presentation at AGU, Baltimore, MD., 21 May 1987. *EOS, Trans. Am. Geophys. Union* **68**(32): 677

188 W.H. Munk and D.M. Farmer. Bringing Physical and Biological Observations into Harmony. In: *Toward a Theory on Biological-Physical Interactions in the World Ocean*, ed. B.J. Rothschild, 549–554. Holland: Kluwer Academic Publishers. Invited paper presented at NATO Advanced Research Workshop, 1–5 June 1987, Château de Bonas, Gers, France

189 W.H. Munk and P.F. Worcester. Ocean Acoustic Tomography *Oceanography* **1**: 8–10

190 W.H. Munk, W.C. O'Reilly, and J.L. Reid. Australia-Bermuda sound transmission experiment (1960) revisited. *J. Phys. Oceanogr.* **18**: 1876–1898

191 Methods for Exploring the Large Scale Ocean Turbulence. In: *Perspectives in Fluid Mechanics*, ed. D. Coles, 1–13. New York: Springer-Verlag. Invited paper presented at Symposium on Perspectives in Fluid Mechanics in honor of Hans-Wolfgang Liepmann, held at California Institute of Technology, Pasadena, California, 10–12 Jan. 1985

1989

192 W. Munk and P. Worcester. An Oceanographic Time Capsule. *Oceanography* **2**: 48–49

[3] *This report rightfully belongs into the 1951–56 Volume of the Collected Papers; it is included at this time because of a renewed interest in the subject.

193 B. Cornuelle, W.H. Munk, and P.F. Worcester. Ocean Acoustic Tomography from Ships. *J. Geophys. Res.* **94**: 6232–6250

194 Munk, W.H., and A.M.G. Forbes. Global Ocean Warming: An Acoustic Measure? *J. Phys. Oceano.* **19**: 1765–1778

195 Global Ocean Warming: Detection by Long-Path Acoustic Travel Time. *Oceanography* **2**(2): 40–41

195a Closing Remarks at the Inaugural Meeting of The Oceanographic Society, Monterey, CA 27–30 August 1989. *Oceanography* **2**(2): 57

195b Acceptance of the 1989 Bowie Medal. San Francisco, CA, 6 December 1989, *EOS, Trans. Am. Geophys. Union* **71**: 13

195c Long term acoustic monitoring of ocean temperature. Invited paper presented at: CCCO, held in Halifax, Nova Scotia, 14–20 Jun. 1989. SCOR-IOC/CCCO-X/3, 14–15

196 R. Davis, M. Gregg, J. Katz, W. Munk, D. Nelson, F. Perkins and J. Vesecky. Sea Surface Films. JASON Tech. Rpt. No. JSR-86-105

1990

197 Greenland Sea Project Group. A Venture Toward Improved Understanding of the Oceans' Role in Climate. *EOS*, 12 Jun. 1990

198 W. Munk, R. Revelle, P. Worcester, and M. Zumberge. Strategy for future measurements of sea level. *National Academy of Sciences Press*, 1990, 221–227

199 The Heard Island Experiment. Indian Inaugural Lecture, including questions and answers in Paris. International Science Lecture Series, Nov. 1990. *National Academy of Sciences Press*

1991

199a Citation for the Ewing Medal. Presented to C. Wunsch, 5 Dec. 1990, AGU, San Francisco, CA. *EOS*, 12 Feb. 1991, 70

200 Refraction of Acoustic Modes in very long-range transmissions. In: *Ocean Variability and Acoustic Propagation*, eds. J. Potter and A. Warn-Varnas, 539–543

200a Dinner Talk at the Symposium on Tactical Oceanography, Naval Postgraduate School, Monterey, California, 13 Mar. 1990. (Unpublished)

201 Refraction of Sound Waves at Polar Latitude. *J. Geophys. Res.* **96**(4): 7015–7022

202 Heard Island Principals. The Heard Island Experiment. *Oceanus* **34**(1): 6–8

203 Global ocean warming. *Sea Wind, Bulletin of Ocean Voice and International Marine Life Alliance* **5**(2): 25–27

204 W. Munk and F. Zachariasen. Refraction of Sound by Islands and Seamounts. *J. Atmospheric and Oceanic Technology* **8**(4): 554–574

205 Bermuda Shadow. Presented at The Robert W. Stewart Symposium, Victoria, BC, 25–26 May 1990. *Atmosphere-Ocean* **29**(2): 183–196

205a Scripps Institution of Oceanography. The Revelle Years. *University of California, San Diego, Scripps Institution of Oceanography, Annual Report*

206 The Heard Island Experiment. Indian Inaugural Lecture. International Science Lecture Series, Nov. 1990. *Naval Research Reviews*. Office of Naval Research. One/1991. Vol. XLII. 2–22

207 An Ode to Southern Swell. Invited paper, Strategies for Future Climate Research Symposium, in honor of Klaus Hasselmann's 60th Birthday. Hamburg, Germany, 14–16 Oct., 1991. Strategies for Future Climate Research, Max-Planck-Institute für Meteorologie, 31–44

208 The Fifteenth Doherty Lecture in Oceans Policy. Invited talk presented at the Center for Oceans Law and Policy. University of Virginia Law School 9 May 1991. *Mar. Technol. Soc. J.* **25**(3): 52–57

208a D.J. Baker and W. Munk. Roger Revelle, Oceanographer, 1909–1991. *Oceanography* **4**(2): 90–91

208b S. Lakoff, W. Munk and S.J. Singer. Roger R. Revelle. In Memoriam. *University of California, Academic Senate*, 159–169

1992

209 Global ocean warming. *Canadian Biodiversity* **1**(4): 10–12

209a Henry Stommel. *Ocean Magazine* **35**: 24–27

209b Ground Breaking for IGPP-II. 5 Jun. 1992. (Unpublished)

209c On the Occasion of Receiving the UCSD Alumnus Award. 13 Jun. 1992. (Unpublished)

210 A. Baggeroer, and W. Munk. The Heard Island Feasibility Test. *Physics Today* **45**(9): 22–30

211 R. Dashen and W. Munk. A model of ocean noise. *JASON*, Summer Study, 1991. (Unpublished)

211a W. Munk and E. Frieman. Let Roger Revelle Speak for Himself. *Oceanography* **5**(2): 125

212 D. Sheres and W. Munk. On John Isaacs' 1952 Measurements of Surface-Ship and Submarine Wakes. *Oceanography* **5**(2): 121–123

213 Dzieciuch, M., A. Forbes, and W. Munk. Interpretation of GPS Off-sets from a steady course. *J. Atmos. and Oceanic Tech.* **9**(6): 862–866

213a A. Baggeroer and W. Munk. Reply to John Spiesberger. *Physics Today*. (Unpublished. Spiesberger withdrew his reply at the last minute.)

1993

213b Acceptance of the Vetlesen Prize. Columbia University, Jan. 21, 1993. (Unpublished)

213c Acceptance of the ARCS 1993 Scientist of the Year Award. 9 Mar. 1993. (Unpublished)

213d The Ocean "Bible": Reminiscences. *Oceanography* **5**(3): 155–157

214 A. Baggeroer and W. Munk. The Heard Island Feasibiliy Test. *Parity* **08**(04): 4–15

214a Acceptance of the Walter Munk Award. Office of Naval Research and The Oceanography Society. Seattle, WA, 14 Apr. 1993. (Unpublished)

214b Measuring Ocean warming. Invited paper, Oceans, Climate and Man. Proceedings of the International Conference held in Turin, Italy, 15–17 Apr. 1991. (Unpublished)
215a Response of Acoustic Rays and Modes to Surface Warming. *ATOC Occasional Note* **2** (Mar., 1993) (Unpublished)
215b It's the Shear Velocity, Stupid! *ATOC Occasional Note* **8** (May, 1993) (Unpublished)
215c Acoustic Thermometry of Ocean Climate. Special Briefing to Vice-President Al Gore, Washington, DC, 15 Sep. 1993. (Unpublished)
215d It's the Slope Stupid. *ATOC Occasional Note* **13** (September 1993) (Unpublished)
216 The Sound of Oceans Warming. *The Sciences* **33**(5):20–26
217 P.F. Worcester, J.F. Lynch, W.M.L. Morawitz, R. Pawlowicz, P.J. Sutton, B.D. Cornuelle, O.M., Johannessen, W.H. Munk, W.B. Owens, R. Shuchman and R.C. Spindel. Evolution of the large-scale temperature field in the Greenland Sea during 1988–89 from tomographic measurements. *Geophys. Res. Lett.* **20**(20): 2211–2214

1994

218 Heard Island and Beyond. *APS News* "Physics News in 1993" **3**: 5 S2
219 *Chapter 2 of Oceanography: Megascience: The OECD Forum*. Tokyo, 28–30 Sep. 1993. Organization for Economic Co-Operation and Development, OECD 1994, Paris, France
220 W. Munk, A. Baggeroer. The Heard Island papers: A contribution to global acoustics. *J. Acoust. Soc. Am.* **96**(4): 2327–2329
221 W. Munk, A. Baggeroer, R. Spindel, and T. Birdsall. The Heard Island Feasibility Test. *J. Acoust. Soc. Am.* **96**(4): 2330–2342
222 M. Dzieciuch and W. Munk. Differential Doppler as a Diagnostic. *J. Acoust. Soc. Am.* **96**(4): 2414–2424
223 A. Forbes and W. Munk. Doppler-inferred launch angles of global acoustic ray paths. *J. Acoust. Soc. Am.* **96**(4): 2425–2427
224 Acoustic Remote Sensing. The Polar Oceans and Their Role in Shaping The Global Environment. Geophysical Monograph. *Am. Geoph. Union*. **85**: 521–523

1995

224a How to Resolve Modes with Sparse Arrays. *ATOC Occasional Note* **29** (Mar., 1995)
225 W. Munk, P. Worcester, and C. Wunsch. *Ocean Acoustic Tomography*. Cambridge University Press, 450 pp
225a Acoustic Monitoring of the Global Ocean Climate. Conference Proceedings from 1995 SERDP Symposium, 12–14 Apr., 1995. page CEG4
225b Citation for Ewing Medal on John Orcutt. *EOS* **76**: 148–149

226 ATOC Instrumentation Group[4], "Instrumentation for the Acoustic Thermometry of Ocean Climate (ATOC) Prototype Pacific Ocean Network," OCEANS '95, San Diego, CA Oct 9–12, 1995, 1483–1500.

1996

227 J.L. Colosi, B.D. Cornuelle, M.A. Dzieciueh, K. Heany, W.H. Munk, P.F. Worcester, B.D. Dushaw, B.M. Howe, J.A. Mercer, R.I. Odom, R.C. Spindel, A.B. Baggeroer, T. Birdsall, K. Metzger, G. Bold, S. Tan, C. Tindle, M. Guthrie. "A test of acoustic thermometry at ocean basin seales," Poster at 1996 AGU Ocean Sciences Meeting, San Diego, CA. Feb. 12–16, 1996

1997

228a W. Munk. Global Ocean Warming. Science with a Human Face In Honor of Roger Randall Revelle, Harvard University Press, Dorfman R., Rogers P.P. (eds.) 19–23

229 W. Munk. Once again: once again – tidal friction. *Progress in Oceanography* **40**: 7–35

230 W. Munk and C. Wunsch. The Moon, of course... *Oceanography* **10**: 132–134

231b W.H.Munk. *Sampling the oceans from above and beneath.* IAPSO President's Invited Lecture, Melbourne, Australia

240a W.H. Munk, P.F. Worcester, J.A. Mercer, and R.C. Spindel *Ocean Acoustic Observatories. Ocean Acoustics Program Summary: Fiscal Year 1996.* (Office of Naval Research, Arlington, VA) pp. D35–D37

1998

228 Larson, G.D., P.H. Rogers, and W. H. Munk. State Switched Transducers: a new approach to high power, low frequency, underwater projectors. *J. Acoust. Soc. Am* **103**: 1428–1441

228c T.F. Malone, E.D. Goldberg, and W. H. Munk. Roger Randall Dougan Revelle 1909–1991. *Biographical Memoirs National Academy of Sciences* **75**: 1–23

231 W. Munk and C. Wunsch. Abyssal Recipes II: energetics of tidal and wind mixing. *Deep-Sea Research I* **45**: 1977–2010

232 The ATOC Consortium[5]. Ocean Climate Change by Acoustic Tomography, Satellite Altimetry, and Modeling. *Science* **281**: 1327–1332

[4] Howe, B.M., S.G. Anderson, A.B. Baggeroer, J.A. Colosi, K.R. Hardy, D. Horwitt, F.W. Karig, S. Leach, J.A. Mercer, K. Metzger, W. Munk, L.O. Olson, D.A. Peckham, D.A. Reddaway, R.R. Ryan, R.P. Stein, K. von der Heydt, J.D. Watson, S.L. Weslander and P. Worcester

[5] A.B. Baggeroer (MIT), T.G. Birdsall (U. Michigan), C. Clark (Cornell Univ.), J.A. Colosi (WHOI), B.D. Cornuelle, (SIO), D. Costa (UC Santa Cruz), B.D. Dushaw (U. Washington), M. Dzieciuch (SIO), A.M.G. Forbes, (CSIRO, Hobart), C. Hill, B.M. Howe (U. Washington), J. Marshall, D. Menemenlis, J.A. Mercer, K. Metzger (U. Michigan) W. Munk (SIO), R.C. Spindel (U. Washington), D. Stammer, P.F. Worcester (SIO), and C. Wunsch

1999

233 P.F. Worcester, B.D. Cornuelle, M.A. Dzieciuch, W.H. Munk, B. M. Howe, J.A. Mercer, R.C. Spindel, J.A. Colosi, WHOI, K. Metzger, T.G. Birdsall, and A.B. Baggeroer. A test of basin-scale acoustic thermometry using a large-aperture vertical array at 3250-km range in the eastern North Pacific Ocean. *J. Acoust. Soc. Am.* **105**: 3185–3201

234 The ATOC Group[6]. Comparisons of measured and predicted acoustic fluctuations for a 3250-km propagation experiment in the eastern North Pacific Ocean. *J. Acoust. Soc. Am.* **105**: 3202–3218

235 K.W. Fischer, S. Legg, W.H. Munk, R.A. Shuchman, R.W. Garwood, J.P. Palshook. Modeled Radar Surface Signature of Deep Ocean Convection. *IEEE Transactions of Geoscience and Remote Sensing* **37**: 2050–2067

236 J.A. Colosi, the ATOC Group[7]. A Review of Recent Results on Ocean Acoustic Wave Propagation in Random Media: Basin Scales. *IEEE J. of Ocean Engineering* **24**: 138–155

237a W. Munk. How Deep is the Ocean, How High is the Sky, *UCSD Millennium Lecture Series*, University of California at San Diego, 02 Dec. 1999. (Unpublished)

2000

237 W. Munk. The Luck of Walter Munk. *Kyoto Prizes & Inamori Grants 1999.* 127–153

237b W. Munk. Oceanography before, and after, the advent of Satellites. In *Satellites, Oceanography and Society,* ed. David Halpern, Elsevier: 1–4

237c W. Munk. Achievements in Physical Oceanography. *50 Years of Ocean Discovery, National Academy Press*: 44–50 and following p. 90

237d W. Munk. National Priorities and the Creation of a Discipline. *The Oceanography Greats Colloquia Series*, Scripps Institution of Oceanography, La Jolla, Ca. (Unpublished)

237e W. Munk. Solving the Tidal Problem: David E. Cartwright, 'Tides: A Scientific History.' *Notes Rec. R. Soc. Lond.* **54**(1): 116–119

238 W. Munk, L. Armi, K. Fischer, and F. Zachariasen. Spirals on the Sea. *Proc. R. Soc. Lond.* **456**: 1217–1280.

238a J. Orcutt, E. Bernard, C. Chiu, C. Collins, C. deGroot-Hedlin, R. Dziak, C. Fox, W. Hodgkiss, W. Kuperman, J. Mercer, W. Munk, R. Odom, M. Park, D. Soukup, R. Spindel, F. Vernon, and P. Worcester (2000) Long-Term

[6] John A. Colosi and Edward Scheer, WHOI; Stanley Flatte, UC Santa Cruz; Bruce D. Cornuelle, Matthew A. Dzieciuch, Walter H. Munk and Peter F. Worcester of SIO; Bruce M. Howe, James A. Mercer and Robert C. Spindel of Appl. Phys. Lab. U. Washington; Kurt Metzger and Theodore G. Birdsall, U. Michigan; Arthur B. Baggeroer, MIT

[7] A.B. Baggeroer (MIT), T.G. Birdsall (U. Michigan), C. Clark (Cornell Univ.), J.A. Colosi (WHOI), B.D. Cornuelle, (SIO), D. Costa (UC Santa Cruz), B.D. Dushaw (U. Washington), M. Dzieciuch (SIO), A.M.G. Forbes, (CSIRO, Hobart), B.M. Howe (U. Washington), D. Menemenlis, J.A. Mercer, K. Metzger (U. Michigan) W. Munk (SIO), R.C. Spindel (U. Washington), P.F. Worcester (SIO), and C. Wunsch

Fig. B.1 Walter and Larry Armi searching for ocean spirals, with Judith wearing her traditional straw hat (2001). Drawing by Mike Dormer. Spirals on the sea were discovered on the early Apollo Mission in 1969, but had never been satisfactorily explained. Thirty years later, Walter and Larry took a stab at discussing the spiral theory in terms of horizontal sheer instability, modified by rotation

Observations in Acoustics – the Ocean Acoustic Observatory Federation. *Oceanography.* **13**(2): 57–63

238d W. Munk. Listening to Ocean Climate. *Reimar Lust Lecture.* MPI, Hamburg, 31 Mar. 2000

238e P. Worcester, B.D. Dushaw, The ATOC Group[8]. A comparison of acoustic thermometry, satellite altimetry, and other observations of ocean tempera-

[8] A.B. Baggeroer (MIT), T.G. Birdsall (U. Michigan), C. Clark (Cornell Univ.), J.A. Colosi (WHOI), B.D. Cornuelle, (SIO), D. Costa (UC Santa Cruz), B.D. Dushaw (U. Washington), M. Dzieciuch (SIO), A.M.G. Forbes, (CSIRO, Hobart), B.M. Howe (U. Washington), D. Menemenlis, J.A. Mercer, K. Metzger (U. Michigan) W. Munk (SIO), R.C. Spindel (U. Washington), P.F. Worcester (SIO), and C. Wunsch

ture in the North Pacific Ocean. *PORSEC (The Fifth Pacific Ocean Remote Sensing Conference)* Proceedings. I: 1–4, Goa, India, 05 Dec. 2000

2001

238b G. Bold, C. Chiu, J. Colosi, B. Cornuelle, B. Dushaw, Y. Desaubies, M. Dzieciuch, A. Forbes, F. Gaillard, J. Gould, B. Howe, M. Lawrence, J. Lynch, D. Menemenlis, J. Mercer, P. Mikhalevsky, W. Munk, I. Nakano, F. Schoot, U. Send, R. Spindel, T. Terre, P. Worcester, C. Wunsch. Observing the Ocean in the 2000s: A Strategy for the Role of Acoustic Tomography in Ocean Climate Observation. *Proc. of the Intl. Conf. on the Ocean Observing System for Climate.* (OCEANOBS99): 391–418, St. Raphael, France, 18–22 Oct. 1999

238c W. Munk. Harald Ulrik Sverdrup (1888–1957): celebrating the return of the MAUD 75 years ago. H.U. Sverdrup Symposium, Oslo, Norway, 21 Sep. 2000. *Polar Research* **20**(2): 129–138

239 M. Dzieciuch, P. Worcester, and W. Munk. Turning point filters: Analysis of sound propagation on a gyre-scale. *J. Acoust. Soc. Am.* **110**(1): 135–149

239a Townes, C. and W. Munk. William Aaron Nierenberg. *Physics Today*. **54**(6): 74–75

239b W. Munk. Spirals on the Sea. *Sci. Mar.* **65**(2): 193–198. 36th European Marine Biology Symposium: *A Marine Science Odyssey into the 21st Century*, Maó (Menorca), Spain, 17–22 Sep. 2001

239c W. Munk and L. Armi. Spirals on the Sea: A Manifestation of Upper-Ocean Stirring. 12th *Proceedings 'Aha Huliko'a: From Stirring to Mixing in a Stratified Ocean*: 81–86

239d W. Munk. Internal Tidal Mixing. In: *Encyclopedia of Ocean Sciences*, ed. J.H. Steele, San Diego, CA: Academic Press: 1323–1327

2002

240 W. Munk, M. Dzieciuch, and S. Jayne. Millennial Climate Variability; Is There a Tidal Connection? *J. Climate* **15**: 370–385

240a W.H. Munk, P.F. Worcester, J.A. Mercer, and R.C. Spindel. Ocean Acoustic Observatories. Ocean Acoustics Program Summary: Fiscal Year 1996. (Office of Naval Research, Arlington, VA) pp. D35–D37

241 W. Munk. The Evolution of Physical Oceanography in the Last Hundred Years. *Oceanography* **15**(1): 135–141. The Joint IAPSO/IABO Assembly 2001: *An Ocean Odyssey*, Mar del Plata, Argentina

242 W. Munk. Twentieth century sea level: an enigma. *Proc. Natl. Ac. Sc.* **99**: 6550–6555

242a W. Munk. What Does a Bunch of Sailors Know about Starting a University? *Chronicles* (UCSD Emeriti Association) **Feb. 2002**: 3–4

242b W. Munk. Testimony. *The U.S. Commission on Ocean Policy, Southwest Regional Meeting*, Cabrillo Marine Aquarium, San Pedro, California, 18 April 2002. (Unpublished)

242c R. Davis, D. Day, E. Frieman, E. Goldberg, and W. Munk. William A. Nierenberg. *In Memoriam 2001*: 179–182

242d W. Munk. Acoustic Tomography. In The *Encyclopedia of Global Environmental Change* **1**: 161, Chichester: John Wiley & Sons, Ltd

243 W. Munk and D. Day. Harald U. Sverdrup and the War Years. *Oceanography* **15**(4): 7–29

243a W. Munk. Harald Ulrik Sverdrup. In The *Encyclopedia of Global Environmental Change* **1**: 708, Chichester: John Wiley & Sons, Ltd

243b W. Munk. Origin of the "Ocean Bible." American Geophysical Union 2002 Fall Meeting, San Francisco, California, 08 Dec. 2002

243c W. Munk. The Sverdrup Years: A Personal Recollection. In: *Oceanographic History; The Pacific and Beyond*, eds. K.R. Benson and P.F. Rehbock, pp. 30–31, University of Washington Press. *Invited Talk – Proceedings of the 5thInternational Congress of History of Oceanography*, Scripps Institution of Oceanography, La Jolla, California, 7–14 Jul., 1993

2003

243d W. Munk and D. Day. Harald U. Sverdrup. In: *Coming of Age: Scripps Institution of Oceanography*, ed. R.L. Fisher, E.D. Goldberg, and C.S. Cox, pp. 177–201, Scripps Institution of Oceanography, University of California, San Diego

244 W. Munk. Ocean Freshening, Sea Level Rising. *Science* **300**: 2041–2043

245 W.H. Munk, P.F. Worcester, and R.C. Spindel. Acoustic Thermometry of Ocean Climate. Preface to *Collected Papers, Acoustic Thermometry of Ocean Climate (ATOC)*

245a W. Munk, F. Gilbert, J. Orcutt, M. Zumberge, and R. Parker. The Cecil H. and Ida M. Green Institute of Geophysics and Planetary Physics (IGPP). *Oceanography* **16**(3): 34–44

245b A.E. Maxwell, W.H. Munk, F. Gilbert, and C.S. Cox. Sir Edward Bullard. In: *Coming of Age: Scripps Institution of Oceanography*, ed. R.L. Fisher, E.D. Goldberg, and C.S. Cox, pp. 13–20, Scripps Institution of Oceanography, University of California, San Diego

2004

245c P.F. Worcester, and W. Munk. Human-generated Ocean Noise and the Effects on Marine Life: Commentary on Ocean Acoustic Tomography. *MTS Journal* **37**(4): 78–82

245d W. Munk, N. Oreskes, and R. Muller. Gordon James Fraser MacDonald. *Biographical Memoirs: National Academy of Science* **84**: 2–26

246 P. Wadhams, and W. Munk. Ocean Freshening, Sea Level Rising, Sea Ice Melting. *Geophys. Res. Lett.* **31**(11): L11311, 10.1029/2004GL020039

247 W. Munk and D. Day. IVY-MIKE. *Oceanography* **17**(2): 96–105

248 M. Dzieciuch, W. Munk, and D.L. Rudnick. The propagation of sound through a spicy ocean; the SOFAR overture. *J. Acoust. Soc. Am.* **116**(3): 1447–1462

248a C.F. Kennel, R.S. Lindzen, and W. Munk. William Aaron Nierenberg. *Biographical Memoirs: National Academy of Science* **85**: 1–20

249 P. Roux, W.A. Kuperman, and The NPAL Group[9]. Extracting coherent wave fronts from acoustic ambient noise in the ocean. *J. Acoust. Soc. Am.* **116**: 1995–2003

249a B.M. Howe, B.D. Cornuelle, B.D. Dushaw, M.A. Dzieciuch, D. Menemenlis, J.A. Mercer, W.H. Munk, R.C. Spindel, D. Stammer, P.F. Worcester, and M. Zarnetske. Acoustic remote sensing of large-scale temperature variability in the North Pacific Ocean. *OCEANS'04, MTS/IEEE Techno-Ocean '04*, Kobe, Japan, 1504–1506

2005

250 J.A. Colosi, A.B. Baggeroer, B.D. Cornuelle, M.A. Dzieciuch, W.H. Munk, P.F. Worcester, B.D. Dushaw, B.M. Howe, J.A. Mercer, R.C. Spindel, T.G. Birdsall, K. Metzger, and A.M.G. Forbes. Analysis of multipath acoustic field variability and coherence in the finale of broadband basin-scale transmissions in the North Pacific Ocean. *J. Acoust. Soc. Am.* **117**: 1538–1564

251 R.K. Andrew, B.M. Howe, J.A. Mercer, and The NPAL Group[10]. Transverse horizontal spatial coherence of deep arrivals at megameter ranges. *J. Acoust. Soc. Am.* **117**: 1511–1526

252 A.G. Voronovich, V.E. Ostashev, and The NPAL Group[11]. Horizontal refraction of acoustic signals retrieved from the North Pacific Acoustic Laboratory billboard array data. *J. Acoust. Soc. Am.* **117**: 1527–1537

253 M.D. Vera, K.D. Heaney, and The NPAL Group [12]. The effect of bottom interaction on transmissions from the North Pacific Acoustic Laboratory Kauai source. *J. Acoust. Soc. Am.* **117**: 1624–1634

254 A.B. Baggeroer, E.K. Scheer, and The NPAL Group[13]. Statistics and vertical directionality of low-frequency ambient noise at the North Pacific Acoustic Laboratory site. *J. Acoust. Soc. Am.* **117**: 1643–1665

254a W. Munk. Let Roger speak for himself. *Chronicles* (UCSD Emeriti Association) **Mar. 2005**: 3–4

254b W. Munk, F. Spiess, F. Seible. Keeping Lindbergh Afloat: Another View on San Diego's Airport Dilemma. *Voice of San Diego*. 5 Jul. 2005, http://www.voiceofsandiego.org/site/apps/s/content.asp?c=euLTJbMUKvH&b=291837&ct=1146381

---

[9] J.A. Colosi, B.D. Cornuelle, B.D. Dushaw, M.A. Dzieciuch, B.M. Howe, J.A. Mercer, W. Munk, R.C. Spindel, and P.F. Worcester

[10] J.A. Colosi, B.D. Cornuelle, B.D. Dushaw, M.A. Dzieciuch, B.M. Howe, J.A. Mercer, W.H. Munk, R.C. Spindel, and P.F. Worcester

[11] J.A. Colosi, B.D. Cornuelle, B.D. Dushaw, M.A. Dzieciuch, B.M. Howe, J.A. Mercer, W.H. Munk, R.C. Spindel, and P.F. Worcester

[12] J.A. Colosi, B.D. Cornuelle, B.D. Dushaw, M.A. Dzieciuch, B.M. Howe, J.A. Mercer, W.H. Munk, R.C. Spindel, and P.F. Worcester

[13] J.A. Colosi, B.D. Cornuelle, B.D. Dushaw, M.A. Dzieciuch, B.M. Howe, J.A. Mercer, W.H. Munk, R.C. Spindel, and P.F. Worcester

255 P.F. Worcester, W.H. Munk, and R.C. Spindel. Acoustic remote sensing of ocean gyres. *Acoustics Today* **1**: 11–17

255a W. Munk. WHOI. *WHOI 75th Anniversary Symposium*. 21 Sep. 2005

255b W. Munk. Tides and the Climate. *Keeling Forum*, 07 Oct. 2005

255c W. Munk. Foreward. In: *HYDRO TO NAVOCEANO: 175 Years of Ocean Survey and Prediction by the U.S. Navy, 1830–2005,* C.C. Bates, pp. xv–xvi, Rockton, IL: Corn Field Press

2006

255d W. Munk. Ocean Acoustic Tomography: From a stomy start to an uncertain future. In *History of Oceanography*. ed. M. Jochum and R. Murtugudde, pp. 119–138, New York, NY: Springer

255e W. Munk. *Sounds in the Sea* Review. *Oceanography* 19(1): 192–194

255f W. Munk. From Ocean Acoustics to Acoustic Oceanography. *EOS Transactions AGU* 87: Ocean Sciences Meeting Supplement, Abstract OS51A-02

255g W. Munk and D. Rudnick. Penetrating the Deep Sound Channel; A Geometric Measure. *EOS Transactions AGU* **87**: Ocean Sciences Meeting Supplement, Abstract OS52J-02

255h M.A. Dzieciuch, P.F. Worcester, W.H. Munk, D.L. Rudnick, and L.J. Van Uffelen. Propagation of sound through a spicy ocean. *EOS Transactions AGU* **87**: Ocean Sciences Meeting Supplement, Abstract OS52J-03

256 D.L. Rudnick and W. Munk. Scattering from the mixed layer base into the sound shadow. *J. Acous. Soc. Am.* **120**(5): 2580–2594

257 J.A. Colosi and W. Munk. Tales of the Venerable Honolulu Tide Gauge. *J. Phys. Oceanogr.* **36**(6): 967–996

2007

258 W. Munk and B. Bills. Tides and the climate; some speculations. *J. Phys. Ocean.* **37**(2): 135–147

258a M.P. Brenner and contributors (G. Chandler, P. Dimotakis, F. Dyson, J. Goodman, D. Hammer, J. Katz, D. Meiron, A. Mulliken, W. Munk, W. Press, and J. Sullivan). *Navy Ship Underwater Shock Prediction and Testing Capability Study*. Mc Lean, Virginia: JASON (JSR-07-200)

258b D. Day and W. Munk. Revelle, Roger Randall Dougan. In *New Dictionary of Scientific Biography*. 8 vols. Scribner's. 4400 p.

2008

259 W.E. Farrell and W. Munk. What do deep sea pressure fluctuations tell about short surface waves?. *Geophys. Res. Lett.* **35**, L19605, doi:10.1029/2008GL035008

259a W. Munk and D. Day. Glimpses of Oceanography in the Postwar Period. *Oceanography* **21**(3): 16–23

2009

260 W. Munk. An Inconvenient Sea-Truth: Spread, Steepness and Skewness of Surface Slopes. *Ann. Rev. Mar. Sci.* **1**: 377–415, 10.1146/annurev.marine.010908.163940.

260a W. Munk. The William Leighton Jordon, Esq. Award, submitted

260b P.F. Worcester and W.H. Munk. Sound Transmission Through a Fluctuating Ocean (1979) and Ocean Acoustic Tomography (1995): An intertwined history. *J. Acoust. Soc. Am.* **126**(4, Part 2), 158th Meeting of the Acoustical Society of America, San Antonio, Texas, 26–30 October 2009: 2001aAO2001, 2157

261 B.D. Dushaw, P.F. Worcester, W.H. Munk, R.C. Spindel, J.A. Mercer, B.M. Howe, K. Metzger Jr., T.G. Birdsall, R.K. Andrew, M.A. Dzieciuch, B.D. Cornuelle, and D. Menemenlis. A decade of acoustic thermometry in the North Pacific Ocean. *J. Geophys. Res.* **114**: C07021, doi:10.1029/2008JC005124

261a P.F. Worcester and W.H. Munk. The role of acoustics in ocean observing systems. Paper presented at Oceanography in 2025

261b B.D. Dushaw, W.W.L. Au, A. Beszczynska-Möller, R.E. Brainard, B.D. Cornuelle, T.F. Duda, M.A. Dzieciuch, A.M.G. Forbes, L. Freitag, J.-C. Gascard, A.N. Gavrilov, J. Gould, B.M. Howe, S.R. Jayne, O.M. Johannessen, J.F. Lynch, D. Martin, D. Menemenlis, P.N. Mikhalevsky, J.H. Miller, S.E. Moore, W.H. Munk, J. Nystuen, R.I. Odom, J.A. Orcutt, T. Rossby, H. Sagen, S. Sandven, J. Simmen, E. Skarsoulis, B. Southall, K.M. Stafford, R. Stephen, K.J. Vigness Raposa, S. Vinogradov, K.B. Wong, P.F. Worcester, and C. Wunsch. A global ocean acoustic observing network. Paper presented at the Proceedings of OceanObs'09: Sustained Ocean Observations and Information for Society

261c W. Munk. On Roger Revelle. Paper presented at the Roger Revelle 100th Birthday Celebration

261d C. Pendarvis and W. Munk. Where the Swell Begins: An Interview with Walter Munk. *Surfer's J.* **18**(6): 92–99

2010

262 W. Farrell and W. Munk. Booms and Busts in the Deep. *J. Phys. Oceanogr.* in press

262a W. Munk and C. Pendarvis. Where the Swell Begins. Groundswell Publication, submitted

263 W. Munk, H. von Storch, and K. Hasselman. Seventy Years of Exploration in Oceanography. Springer, in press

Index